Communications
in Computer and Information Science   **1511**

More information about this series at https://link.springer.com/bookseries/7899

Abbas M. Al-Bakry · Safaa O. Al-Mamory ·
Mouayad A. Sahib · Loay E. George ·
Jaafar A. Aldhaibani · Haitham S. Hasan ·
George S. Oreku (Eds.)

New Trends in Information and Communications Technology Applications

5th International Conference, NTICT 2021
Baghdad, Iraq, November 17–18, 2021
Proceedings

 Springer

Editors
Abbas M. Al-Bakry
University of Information Technology
and Communications
Baghdad, Iraq

Safaa O. Al-Mamory 🆔
University of Information Technology
and Communications
Baghdad, Iraq

Mouayad A. Sahib
University of Information Technology
and Communications
Baghdad, Iraq

Loay E. George
University of Information Technology
and Communications
Baghdad, Iraq

Jaafar A. Aldhaibani
University of Information Technology
and Communications
Baghdad, Iraq

Haitham S. Hasan
University of Information Technology
and Communications
Baghdad, Iraq

George S. Oreku
Open University of Tanzania
Dar es Salaam, Tanzania

ISSN 1865-0929   ISSN 1865-0937 (electronic)
Communications in Computer and Information Science
ISBN 978-3-030-93416-3   ISBN 978-3-030-93417-0 (eBook)
https://doi.org/10.1007/978-3-030-93417-0

This Springer imprint is published by the registered company Springer Nature Switzerland AG
The registered company address is: Gewerbestrasse 11, 6330 Cham, Switzerland

Preface

The 5th International Conference on New Trends in Information and Communications Technology Applications (NTICT 2021), held in Baghdad, Iraq, during November 17–18, 2021, was hosted and organized by the University of Information Technology and Communications (UoITC). NTICT focuses on specific topics in computer methodologies, networks, and applied computing, and it is the first conference series in Iraq, in its fifth edition, publishing its proceedings in Springer's CCIS series.

The NTICT conference series aims at providing a meeting place for advanced discussion of evolving applications of computer methodologies in the fields of artificial intelligence and machine learning, networks (as a group of computers and other devices connected in some way to be able to exchange data), and applied computing (considered as the intersection of information technology which focuses on technical computing concepts and the development of skills in organizational leadership and business strategy). Thus, it brings both young researchers and senior experts together to disseminate novel findings and encourage practical proficiency in the aforementioned fields. The NTICT conference series also aims to enhance the scientific research movement in Iraq.

NTICT 2021 received a total of 64 local and international submissions, including short and long manuscripts, 12 of which were screened out. The remaining manuscripts were delegated to reviewers according to their research interests. The review and selection procedure was performed carefully, with each manuscript being reviewed by at least three reviewers through a double-blind peer-review process. After reviewing, only 13 full papers were accepted, giving an acceptance rate of about 25%. The accepted papers were distributed into three tracks: Computer Methodologies, Networks, and Applied Computing.

We thank the Program Committee members and reviewers for their time and valuable comments during the reviewing process. The high-quality program of NTICT 2021 would not have been possible without their expertise and dedication. The reviewing process timetable was ambitious, and the reviewers' efforts to submit their reports within the reviewing period were greatly appreciated. Their comments were instructive for us in selecting the papers. We also appreciated the use of the Open Conference System (OCS) software for managing the submissions.

We are grateful to the keynote speakers for their excellent presentations. Thank you! Thanks also go to all authors for submitting their work, the conference participants, and theUniversity of Information Technology and Communications (UoITC) for its continuous support and for making this conference proceedings possible. We hope that readers find the proceedings exciting, and that the papers collected in this volume offer inspiration and open new research windows.

November 2021

Abbas M. Al-Bakry
Safaa O. Al-Mamory
Mouayad A. Sahib
Loay E. George
Jaafar A. Al-Dhaibani
Haitham S. Hasan
George S. Oreku

Organization

General Chair

Abbas M. Al-Bakry   UOITC, Iraq

Co-chair

Hasan Shaker Magdy   Al-Mustaqbal University College, Iraq

Program Committee

Abbas M. Al-Bakry   UOITC, Iraq
Safaa O. Al-Mamory   UOITC, Iraq
Mouayad A. Sahib   UOITC, Iraq
Loay E. George   UOITC, Iraq
George S. Oreku   Open University of Tanzania, Tanzania
Haitham S. Hasan   UOITC, Iraq
Jaafar A. Al-Dhaibani   UOITC, Iraq

Steering Committee

Loay Edwar George   UOITC, Iraq
Jane J. Stephan   UOITC, Iraq
Mahdi N. Jasim   UOITC, Iraq
Ahmed A. Hasim   UOITC, Iraq
Inaam Rikan Hassan   UOITC, Iraq
Ahmed Sabah Ahmed   UOITC, Iraq
Ali Hassan Tarish   UOITC, Iraq
Nagham Hamid Abdul-Mahdi   UOITC, Iraq

International Scientific Committee

Abbas Al-Bakry   UOITC, Iraq
Abdel-Badeeh Salem   Ain Shams University, Egypt
Abdelghani Aissaoui   University Tahri Mohamed of Bechar, Algeria
Abdelnaser Omran   Bright Star University, Libya
Abdulamir Karim   University of Technology, Iraq
Abdulrahman Jasim   Al-Iraqia University, Iraq
Abid Yahya   Botswana International University of Science and
                            Technology, Botswana

Adolfo Guzmán-Arenas	Instiruto Politécnico Nacional, Mexico
Ahmad Bader	Middle Technical University, Iraq
Ahmad Mohammad	Mustansiriyah University, Iraq
Ahmed Hashim	UOITC, Iraq
Ahmed Shamil Mustafa	Al-Maarif University College, Iraq
Aini Syuhada Md Zain	Universiti Malaysia Perlis, Malaysia
Alaa Al-Waisy	Imam Ja'afar Al-Sadiq University, Iraq
Alaa Abdulateef	Universiti Utara Malaysia, Malaysia
Alaa Al-Hamami	British University of Bahrain, Bahrain
Alaa Taqa	University of Mosul, Iraq
Alejandro Zunino	ISISTAN, UNICEN, and CONICET, Argentina
Ali Abbas	Imam Ja'afar Al-Sadiq University, Iraq
Ali Al-Ataby	University of Liverpool, UK
Ali Alnooh	UOITC, Iraq
Ali Al-Shuwaili	UOITC, Iraq
Ali Al-Timemey	University of Baghdad, Iraq
Ali Idrees	University of Babylon, Iraq
Amera Melhum	University of Duhok, Iraq
Athraa Jani	Mustansiriyah University, Iraq
Attakrai Punpukdee	Naresuan University, Thailand
Aws Yonis	Ninevah University, Iraq
Ayad Abbas	University of Technology, Iraq
Azhar Al-Zubidi	Al-Nahrain University, Iraq
Basim Mahmood	University of Mosul, Iraq
Belal Al-Khateeb	University of Anbar, Iraq
Buthainah Abed	UOITC, Iraq
Cik Feresa Mohd Foozy	Universiti Tun Hussein Onn Malaysia, Malaysia
Dena Muhsen	University of Technology, Iraq
Dennis Lupiana	Institute of Finance Management, Tanzania
Kumar	Jawaharlal Nehru Technological University, India
Wafaa Abedi	City University College of Ajman, UAE
Duraid Mohammed	Al-Iraqia University, Iraq
Ehsan Al-Doori	Al Nahrain University, Iraq
Emad Mohammed	Orther Technical University, Iraq
Essa I. Essa	University of Kirkuk, Iraq
Fadhil Mukhlif	University of Technology Malaysia, Malaysia
Farah Jasem	University of Anbar, Iraq
Fawzi Al-Naima	Al-Ma'moon University College, Iraq
George S. Oreku	TIRDO, Tanzania
Ghani Hashim	University of Lorraine, France
Haider Hoomod	Mustansiriyah University, Iraq
Haithm Hasan	UOITC, Iraq
Hamidah Ibrahim	Universiti Putra Malaysia, Malaysia
Hanaa Mahmood	University of Mosul, Iraq
Harith Al-Badrani	Ninevah University, Iraq
Hasan Al-Khaffaf	University of Duhok, Iraq

Haydar Al-Tamimi	University of Technology, Iraq
Hemashree Bordoloi	Assam Don Bosco University, India
Hiba Aleqabie	University of Karbala, Iraq
Hussam Mohammed	University of Anbar, Iraq
Ibtisam Aljazaery	University of Babylon, Iraq
Idress Husien	University of Kirkuk, Iraq
Intisar Al-Mejibli	UOITC, Iraq
Jaafar Aldhaibani	UOITC, Iraq
Jane Stephan	UOITC, Iraq
Junita Mohd Nordin	Universiti Malaysia Perlis, Malaysia
Jyoti Prakash Singh	National Institute of Technology Patna, India
Kiran Sree Pokkuluri	Jawaharlal Nehru Technological University, India
Kuldeep Kumar	National Institute of Technology Jalandhar, India
Kunal Das	West Bengal State University, India
Layla H. Abood	University of Technology, Iraq
Litan Daniela	Hyperion University of Bucharest, Romania
Loay Edwar	UOITC, Iraq
Mafaz Alanezi	University of Mosul, Iraq
Mahdi Abed Salman	University of Babylon, Iraq
Mahdi Jasim	UOITC, Iraq
Malik Alsaedi	Al-Iraqia University, Iraq
Marinela Mircea	Bucharest University of Economic Studies, Romania
Miguel Carriegos	Universidad de León, Spain
Moceheb Shuwandy	Tikrit University, Iraq
Modafar Ati	Abu Dhabi University, UAE
Mohammad Al-Mashhadani	Al-Maarif University College, Iraq
Mohammad Sarfraz	Aligarh Muslim University, India
Mohammed Alkhabet	University Putra Malaysia, Malaysia
Mohammed Aal-Nouman	Al-Nahrain University, Iraq
Mohammed Al-Khafajiy	University of Reading, UK
Mohammed Al-Neama	Mosul University, Iraq
Mouayad Sahib	UOITC, Iraq
Muhammad Raheel Mohyuddin	NCBA&E, Pakistan
Muhammad Zakarya	Abdul Wali Khan University, Pakistan
Muhsen Hammoud	Federal University of ABC, Brazil
Muthana Mahdi	Mustansiriyah University, Iraq
Nada M. Ali	University of Baghdad, Iraq
Nadhir Ibrahim Abdulkhaleq	UOITC, Iraq
Nadia Al-Bakri	Al Nahrain University, Iraq
Noor Maizura Mohamad Noor	Universiti Malaysia Terengganu, Malaysia
Nor Binti Jamaludin	National Defence University of Malaysia, Malaysia
Nur Iksan	Universitas Riau Kepulauan, Indonesia
Omar Abdulrahman	University of Anbar, Iraq
Omar Saleh	Ministry of Higher Education and Scientific Research, Iraq

Omar Salman	Al-Iraqia University, Iraq
Paula Bajdor	Czestochowa University of Technology, Poland
Qais Qassim	University of Technology and Applied Sciences, Oman
Qaysar Mahdy	Tishk International University, Iraq
Raja Azura	Universiti Pendidikan Sultan Idris, Malaysia
Razwan Najimaldeen	Cihan University-Duhok, Iraq
Robert Laramee	University of Nottingham, UK
Ruslan Al-Nuaimi	Al-Nahrain University, Iraq
S. Nagakishore Bhavanam	Acharya Nagarjuna University, India
Saad Dheyab	UOITC, Iraq
Safaa Al-Mamory	UOITC, Iraq
Santhosh Balan	Jawaharlal Nehru Technological University, Hyderabad, India
Sarmad Ibrahim	Mustansiriyah University, Iraq
Sarmad Hadi	Al-Nahrain University, Iraq
Shaheen Abdulkareem	University of Duhok, Iraq
Shaimaa Al-Abaidy	University of Baghdad, Iraq
Shumoos Al-Fahdawi	Al-Maarif University College, Iraq
Siddeeq Ameen	Technical College of Informatics, Iraq
Sk Sarif Hassan	Vidyasagar University, India
Thaker Nayl	UOITC, Iraq
Venkatesh R.	Anna University, India
Walead Sleaman	Tikrit University, Iraq
Yafia Radouane	Ibn Tofail University, Morocco
Yaseen Yaseen	University of Anbar, Iraq
Yousif Hamad	Seberian Federal University, Russia
Yousif Sulaiman	Al-Maarif University College, Iraq
Yusra Al-Irhayim	University of Mosul, Iraq
Zailan Siri	Universiti Malaya, Malaysia
Zainab Ahmed	University of Baghdad, Iraq
Zeyad Karam	Al-Nahrain University, Iraq
Zeyad Younus	University of Mosul, Iraq
Ziad Al-Abbasi	Middle Technical University, Iraq

Secretary Committee

Jwan Knaan Alwan	UOITC, Iraq
Alya Jemma	UOITC, Iraq
Ali Abed Al-Kareem	UOITC, Iraq
Ammar AbdRaba Sakran	UOITC, Iraq
Suhaib S. Al-Shammari	UOITC, Iraq
Ali Jassim Mohammed	UOITC, Iraq

Contents

Computing Methodologies

Semi-automatic Ontology Learning for Twitter Messages Based
on Semantic Feature Extraction 3
   Yasir Abdalhamed Najem and Asaad Sabah Hadi

Audio Compression Using Transform Coding with LZW and Double Shift
Coding 17
   Zainab J. Ahmed and Loay E. George

The Effect of Speech Enhancement Techniques on the Quality of Noisy
Speech Signals 33
   Ahmed H. Y. Al-Noori, Atheel N. AlKhayyat, and Ahmed A. Al-Hammad

Image Steganography Based on DNA Encoding and Bayer Pattern 49
   Elaf Ali Abbood, Rusul Mohammed Neamah, and Qunoot Mustafa Ismail

Panoramic Image Stitching Techniques Based on SURF and Singular
Value Decomposition 63
   Nidhal K. EL Abbadi, Safaa Alwan Al Hassani,
   and Ali Hussein Abdulkhaleq

Lossless EEG Data Compression Using Delta Modulation and Two Types
of Enhanced Adaptive Shift Coders 87
   Hend A. Hadi, Loay E. George, and Enas Kh. Hassan

Automatic Classification of Heart Sounds Utilizing Hybrid Model
of Convolutional Neural Networks 99
   Methaq A. Shyaa, Ayat S. Hasan, Hassan M. Ibrahim,
   and Weam Saadi Hamza

Hybrid Approach for Fall Detection Based on Machine Learning 111
   Aythem Khairi Kareem and Khattab M. Ali Alheeti

Three N-grams Based Language Model for Auto-correction of Speech
Recognition Errors 131
   Imad Qasim Habeeb, Hanan Najm Abdulkhudhur,
   and Zeyad Qasim Al-Zaydi

Networks

A Proposed Dynamic Hybrid-Based Load Balancing Algorithm to Improve
Resources Utilization in SDN Environment 147
    Haeeder Munther Noman and Mahdi Nsaif Jasim

Energy-Saving Adaptive Sampling Mechanism for Patient Health
Monitoring Based IoT Networks 163
    Duaa Abd Alhussein, Ali Kadhum Idrees, and Hassan Harb

ETOP: Energy-Efficient Transmission Optimization Protocol in Sensor
Networks of IoT 176
    Ali Kadhum Idrees, Safaa O. Al-Mamory, Sara Kadhum Idrees,
    and Raphael Couturier

Applied Computing

The Extent of Awareness of Faculty Members at Albaydha University
About the Concept of Educational Technology and Their Attitudes
Towards It 189
    Nayef Ali Saleh Al-Abrat and Mohammed Hasan Ali Al-Abyadh

Author Index 209

Computing Methodologies

Semi-automatic Ontology Learning for Twitter Messages Based on Semantic Feature Extraction

Yasir Abdalhamed Najem$^{(\boxtimes)}$ and Asaad Sabah Hadi

Software Department, University of Babylon, Babylon, Iraq
yasir.abd@student.uobabylon.edu.iq,
asaadsabah@uobabylon.edu.iq

Abstract. The enormous spreading of social network media like Twitter is speeding up the process of sharing information and expressing opinions about global health crises, and important events.

Due to the use of different terms for expressing the same topic in a Twitter post, it becomes difficult to build applications such as retrieval of information by following previous mining methods to find a match between words or sentences. In order to solve this problem, it requires providing the knowledge source that collects many terms which reflect a single meaning, Such as ontology. Ontology is the process of representing the concepts of a specific field such as finance or epidemics, with their characteristics and relationships by dealing with the heterogeneity and complexity of terms.

In this paper, the domain ontology for Twitter's Covid-19 post will be developed by following the notion of semantic web layer cake and discuss the depth of terms and relationships extracted in this domain through a set of measurements, the ontology contains more than 900 single concepts and more than 180 multi-word concepts, which are the most concepts used in Twitter posts with the hashtag Corona epidemic which can be used to find semantic similarities between words and sentences.

Keywords: Ontology learning · Terms extraction · Word sense disambiguation · Concept · Taxonomy · Non-taxonomy · Covid-19

1 Introduction

Since the beginning of the epidemic at the end of 2019, 131 million people have been diagnosed with Covid-19 in all countries of the world. The number of deaths has reached 2.84 million [1]. The massive use of a social networks like Twitter has accelerated the exchange and spread of information, opinions and events about public crises such as Covid-19, especially when imposing quarantine measures in most countries of the world. As the world has become heavily dependent on platforms of social media like Twitter to receive news and express their views, the Twitter platform has a large and wide role in spreading and updating news in real-time.

A. M. Al-Bakry et al. (Eds.): NTICT 2021, CCIS 1511, pp. 3–16, 2021.
https://doi.org/10.1007/978-3-030-93417-0_1

The study of Eysenbach and Chew has shown that Twitter can be used in real-time "science of information" studies and it became a source for posting guidance from authorities of health in response to public concerns [2]. Now with the Covid-19 epidemic, Twitter is used by many government officials around the world as one of the main communicate channels to participate in political events and news related to Covid-19 on a regular basis for the general public [3]. Due to the different terms used to express the same topic in the Twitter post, it becomes difficult to create the applications on it without providing a source of knowledge that unites these terms in a single concept such as ontology and the semantic web.

Many researchers described the ontology as a basis for semantic web applications. Due to the significance of the semantic web and its complete dependence on the ontology, building it automatically is an important and at the same time very difficult task. On the one hand, it requires dealing with the bottleneck of gaining knowledge. On the other hand, it is affected by inhomogeneity, scalability, uncertainty, and low quality of web data. Learning ontology includes extracting knowledge through two main tasks: extracting the concepts (which make up the ontology), and extracting the semantic relationships between them [4].

Ontology learning from texts for a specific domain is a process that involves analyzing those texts and extract terminology and concepts related to the domain and their relationships, then mapping the ontology through ontology representation languages, for example, Resource Description Framework (RDF), Resource Description Framework Schema (RDFS), and Web Ontology Language (OWL). Finally, the built ontology is evaluated [5]. The process of building an ontology can generally be done through one of the following three methods: manual generation; collaborative generation (this type requires human intervention in generating an ontology), and robotic generation (semi). Because learning ontology from texts is an automatic or semi-automatic operation, several approaches have emerged in recent years that attempt to automate the construction of ontologies. The ontology learning systems are based on mining knowledge rather than data mining [6], where the knowledge is represented as concepts and relations that connect those concepts to be in a machine-understandable form. The proposed system aims to generate a specific ontology in Twitter post with the hashtag Covid-19.

The remainder of this research can be outline as follows: part (2) states the relevant work, part (3) introduces the ontology structure, and part (4) explains the evaluation method. Finally, part (5) states the concluding remarks.

2 Related Work

Ontology is a formal identification for sharing the important concepts of a domain, where the official form indicates that the ontology should be understandable by machine and its domain can be shared by the community. Much research in ontology focuses on issues related to building and updating the ontology. OL uses various methods from several fields like Natural Language Processing, Machine Learning, Acquisition of Knowledge, Retrieve Information, AI, and Reasoning [7].

Maedche et al. [8] presented an approach to extract information based on ontology with the aid of machine learning. He also provided a framework for ontology learning which a significant step is in their proposed approach, but this model is still described as a theoretical model. Gómez-Pérez et al. [9] provides an inclusive summary of the many ontology learning projects interested in acquiring knowledge from a set of sources like structured data, text, lexicon, diagrams of relational, knowledge bases, etc. [10] Shamsfard et al. provide a framework for learning the ontology of the semantic web, which includes importing, extracting, trimming, refining, and evaluating ontology. This process provides ontology learning engineers with many coordinated tools for ontology design. Added to the framework, they implemented the Text-To-Onto system that supports learning ontologies from the document, lexicon, or old ontologies. However, they did not mention the way in which the documents related to the field from the network are collected, or how to automatic recognize these documents that are needed in building the ontology [11]. Ah-Hwee Tan and Xing Jiang (2010) they used th3 CRCTOL system to construct an ontology for two cases, first one for terrorism and the second ontology for the sporting events domain. BY comparing CRCTOL with Text-To-Onto and Text2Onto at the component level, the evaluation showed that CRCTOL is better for concepts extraction and relations.

[12] Saeed Al. Tariq Helmy provides the learning of an Arabic ontology from unstructured text based on some of the NLP tools available for Arabic text, using the GATE text analyses system to administer and annotations text [13]. Suresh Da. and Ankur P. proposed a method and tool for learning ontology called DLOL, which creates a knowledge base in Description Logic (DL) SHOQ (D) from a set of non-negative IS-A phrasal statements in English.

The main contributions to this paper is the first work that deals with the process of generating an ontology vocabulary knowledge for Twitter post using the hashtag Covid-19 and using two semantic dictionaries to extract the semantic relationship as a triple (concept, relationship, concept) to be used by our future work to inquire, retrieve information and discover information dissemination. Table 1 illustrates the difference between our proposal and the previous Ontology Learning Systems.

Table 1. A comparison of our proposal with previous ontology learning systems.

Ref.	Input type	Term	Taxonomic & non-taxonomic relations	Axiom	Output domain
Maedche and Staab	Structured and semi-structured data (English)	Statistical	Association rules, formal concept analysis and clustering	No	Not specified in special domain
Shamsfard, M. &Barforoush A	Structured (French)	Starting from a small kernel	Using a hybrid symbolic approach, a combination of logical, linguistic based, template driven and heuristic methods	Yes	Not specified in special domain
Xing J and Ah-Hwee T	Unstructured plain text documents	Statistical and lexico-syntactic methods	A rule-based algorithm	No	Terrorism domain and a sport event domain ontology
Saeed A and Tarek Helmy	Un-structured text (Arabic)	Statistical	Adopted a pattern based algorithm	No	Domain Arabic ontology
Sourish D et al	Structured and semi-structured data (English)	Lexical normalization	Special annotator	No	Not specified in special domain
Our approch	Un-structured Tweets text (English)	Statistical, lexico-syntactic and machine learning methods	Used wordnet, babelnet lexical, and Tree summarize Theory	Yes	Covid-19 pandemic

3   Ontology Construction

The design of our system for building Tweet ontology using the Covid-19 hashtag will be discussed. Figure 1 shows an architecture diagram of the proposed work. The system contain a six stages: (1) collecting domain Tweets, (2) pre-processing the raw data (3) extracting the domain term (4) extracting the concept (5) building taxonomic relationships, and (6) building non-taxonomic relationships.

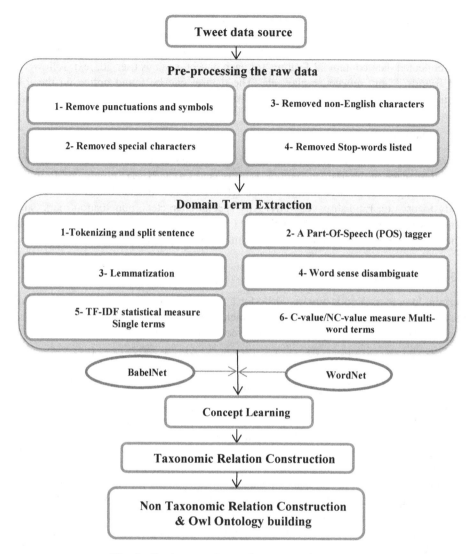

Fig. 1. Semi-automatic ontology generation process

3.1 Domain Tweets Collection

We used the dataset from Kaggle which offers a huge repository of community published data and code. This data is collected by Shane Smith using the Application Programming Interface (API) for Twitter to collect the Tweets that are set From March 29 to April 30, 2020. The data consist of tweets of users who used the following hashtags: #coronavirus, #corona virus outbreak, #corona virus Pandemic, #Covid19,

#Covid_19, #epitwitter, #i have corona, #Stay Home Stay Safe, #Test Trace Isolate. The data consists of about 14,576,636 million tweets posted throughout this period. After eliminating the non-English tweets, a total of 8,584,764 million Tweets remained. The collected data contain 22 features for each Tweet, such as user_id, text, followers, and friends, Lang, retweet, replay, etc. The dataset can be found online on the following website: https://www.kaggle.com/smid80/coronaviruscovid19tweetsearlyApril.

3.2   The Data Pre-processing

As presented in Fig. 1, the Python language is used to clean up the text. At the beginning, each of the hashtag symbol, @users, and link URLs are eliminated from the tweets in the dataset, after which the non-English words are deleted because the present study focuses on English tweets only. Any special characters, punctuation marks and stop words included in [14] from the data are also removed because these do not contribute to the semantic meaning of the message.

3.3   Domain Term Extraction

Terminology extraction is a prerequisite for building the ontology of Tweets. The term may be a single word or multiple words, which refers to a specific meaning in a specific domain. The following steps explain the process of extracting the term:

- Tokenizing and split sentence to reveal sentences and words boundaries.
- Shorthand words to their base by lemmatization.
- Use Part of Speech tagger (POS) to annotate the words with its grammatical category in context, thus specifying if it is a noun, verb, adverb, adjective, etc.
- Word sense disambiguate: In natural language most word is polysemous where they have multiple meaning for example the word "bass" in "it has a bass sound" and "the grilled bass tastes delicious" has a different meaning. When a person reads these two sentences, he knows the meaning of the bass through the words surrounding them, while the machine cannot distinguish the meaning. To remove that ambiguity of the words, a method is used that depends on semantic relation as shown in Table 2. An adaptive lesk algorithm, which relies on the WordNet indicative dictionary, will be used to extract the true meaning of the word. As the algorithm relies on comparing the semantic structure (different levels of hypernym) of the target word with the surrounding words, two words to the right direction and two words to the left direction are taken, and the definition target word is compared to the definitions of those words.

Terminology extraction aims to define the term characteristic of this domain. To extract the related terms, there is a simple mechanism that may refer to concepts in that area by counting the term's frequency in a specific set of tweets as shown in Algorithm (1).

In common, this approach depends on the premise that a term that is repeated in a group of domain-specific texts refers to the concept which belongs to this field. Information retrieval research has shown that there are more active ways to weight the term than simple frequency calculation, and this technique based on weighting scales such as "tf.idf" as shown in Eq. (1), (2), (3), which is used in the present work to extract the individual terms. A total of (918) single-word terms are found.

$$Tf.Idf = log\frac{N}{(df+1)} * If(t,d) \tag{1}$$

$$Idf = log\frac{N}{(df+1)} \tag{2}$$

$$Tf(t,d) = \frac{count\ of\ t\ in\ d}{number\ of\ words\ in\ d} \tag{3}$$

Where 't' is a term, d is a tweet (group of words), N is a number of tweets, tf(t,d) is a number of a term 't' in tweet 'd' / number of words in 'd', and df(t) is the number of times a 't' has appeared in documents.

To find the terms that consist of more than one word, the C-value/NC-value measure is used, which is an important measure for finding multi-word terms that are presented in [15]. It does not depend only on the repetition of the term, but also on its overlap with each other, as shown in Eq. (4). Moreover, this measure also uses contextual clues which are powerful signals of the sequence of some words. A total of (183) multi-word terms were found.

$$C - value = \begin{cases} log_2|a| \cdot f(a) \\ log_2|a| \cdot f(a) - \frac{1}{p(Ta)}\sum_{be(Ta)}f(b) \end{cases} \quad a\ is\ not\ nested \tag{4}$$

'a' is the candidate string, f (.) is the frequency of its appearance in the document; 'Ta' 'Ta' is the group of terms extracted that include a, and (Ta) is the count of those terms.

Table 2. The result of preprocessing and WSD.

Action	Result
Tweet	We took a step to save lives from the early moment of the start of the crisis in China. Now we must take similar action with Europe
Preprocessing sentence	Take step save live early moment start crisis china. must take similar action Europe
WSD	take.v.08 step.v.08 save.n.01 live.v.02 early.a.01 moment.n.06 beginning.n.02 crisis.n.01 china.n.02 remove.v.01 alike.a.01 natural_process.n.01 eu-rope.n.01

Algorithm (1) Domain Term Extraction

1. Input: clean tweet text   // text contains users tweet after pre   -processing
2. Output: Single term, Multi -term // contain the most important single term & multi -word term in a tweet.
3. Begin
4. Read tweet text
5. Tokenizing and a s entence tweet splitter
6. Find Lemmatization of each Word
7. Find part-of-speech (POS) tagger for each word
8. Eliminate ambiguities from words
9. Find most important single term by using equation (1)
10. Find most important multi -word term by using eq uation (4)
11. END

3.4   Concept Learning

In this part, the learning concepts are offered through the use of lexical knowledge bases to extract semantic concepts of the term (the intentional concept description in the form of natural language description) and synonyms of the term. This was done through the use of two semantic dictionaries as shown in Algorithm (2):

- WordNet: It is a lexical online database set up under the Supervised of Dr George Miller in the Laboratory of Cognitive Science at the University of Princeton [16]. WordNet is the more significant resource available to researchers in Linguistics Computational, text analyzes, and numerous related fields. Its layout is inspired by actual lingual, psychological, and computational theories lexical memory of human. The English nouns, verbs, adjectives, and adverbs are regular into synonym group; each one represents a basic lexical concept. Various relationships link sets of synonyms including hypernyms, antonyms, hyponymy, meronyms, holonymy, synonymy, troponymy etc.
- BabelNet is a multi-lingual encyclopedic dictionary with wide encyclopedic and lexicographic covering of terms, as well as a semantic network which relates named entities and concepts in a very large network of semantic relations. It has about 13 million Named Entities. Babel Net is designed within the Sapienza NLP collection and servicing by Babelscape. BabelNet simulates the WordNet where it based on the notion of synset, but it expanded to contain multilingual lexicalizations. Each synset group of BabelNet represents a specific meaning and includes all synonyms that express that meaning in a group of several languages [17].

The Concept will be generated by adding uniquely a textual definition called gloss with their synonyms extracted from above lexical knowledge bases. For example, for the term 'Covid-19', the synonym list would consist of the words 'corona', 'Covid-19', 'coronavirus disease 2019' 'Wuhan coronavirus', '2019-nCoV acute respiratory disease', 'Wuhan virus', 'Wuhan flu', 'novel coronavirus pneumonia', 'coronavirus', 'Wuhan pneumonia'. The official definition of the term would be "a viral disease caused by SARS-CoV-2 that caused a global pandemic in 2020" [17].

Algorithm (2) Concept Learning
1. Input: singl e term, multi-word term   //the term extract by Algorithm. (1)
2. Output: Concept.
3. Begin
4. Read term
5. Fetch the official definition from semantic dictionary (WordNet, BabelNet)
6. Fetch the Synonyms from semantic dictionary (WordNet, BabelNet)
7. Build Concept from term, official definition, Synonyms.
8. END

3.5   Taxonomic Relation Construction

Taxonomy or concept hierarchy is an important part of ontology. Hierarchy relations
are relations which provide tree visions of the ontology and determine the inheriting
between concepts. The WordNet is used, which is organized hierarchically lexical
system motivated by actual linguistic, psychological, and computational theories lexical memory of human. WordNet is more like a thesaurus than a dictionary because it
organizes lexical information in terms of word meanings (or senses) instead of word
forms. The Hypernyms, or the IS-a relation is extracted as a triple (concept, IS-a,
concept) to generate the tree, A can be defined as "a hypernym of B if all B is a (kind
of) A". All levels of the hypernym word can be organize into a hierarchy in which the
sensing are structured from the most specified at the low levels to the most generic
meaning at the top. For example, "motor vehicle" is a hypernym of "self-propelled
vehicle", after that the tree will be compressed by removing parent nodes with less than
2 children [18] as shown in Fig. 2. In case there are three terms (car, Train and Cargo),
and after extracting the hierarchical relationships (IS-a), a tree is constructed for each
term. The sub-trees are combined to build a hierarchical tree that includes all the terms,
and then the tree is summarized to find the final hierarchical tree, as shown in Algorithm (3).

3.6   Non Taxonomic Relation Construction

To obtain nonhierarchical relations such as Synonyms, antonym, meronymy, the API is
used to collect the above relation from BabelNet and WordNet. To exemplify, the term
car has the synonyms (car, auto, automobile, machine, motorcar) and the meronyms
(accelerator, air bag, automobile engine, car horn, car door, car seat, etc.). The definitions of the OWL ontology will be created based on the meta-data above by using
Protégé 5.5.1, as shown in Figs. 3 and 4.

Protégé [19] is a knowledge base and ontology editor generated by University of
Stanford. The Protégé is considered to be a programming language that enables the
creation of domain ontologies and arrange data entry forms for data entry, where it is

used to create the classes, class hierarchies, variable-value restrictions, variables, and the properties of the relationships between classes. The Protégé is a powerful building tools for knowledge engineers and domain ontology specialists.

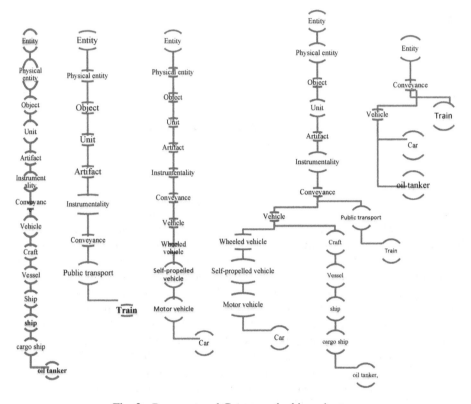

Fig. 2. Represent and Compress the hierarchy tree

4 Evaluation

The ontology evaluation is an important part of learning ontology. It is considered to be a challenging task as there is no standard evaluation available, as in Retrieval of Information where the precision and recall measure is used. The reason can be traced back to the nature of unsupervised of ontology learning method which makes the evaluation harder than supervised learning like classification. Therefore, the approaches of Ontology evaluation can be classified based on several factors like comparing ontologies, usage and application of ontologies, or human evaluation, in order to meet ontology requirements and compatibilities.

The present work is based on human evaluation by using OntoMetrics [20] which is a web-based tool product by Germany Rostock University that validates and displays statistics about a given ontology. The ontology is evaluated through three schema metrics as shown in Table 3.

4.1 Inheritage Richness

It is a metric that describes the information distribution of information over various levels of the inheritage tree of ontology or the prevalence of parental classes. It is a proper indication of how good knowledge is classified into various classes and ontology subcategories. This scale is distinguished by its ability to distinguish the horizontal ontology, for example that color contains a large number of direct branches (Red, Green, Blue, etc.), and while in the vertical ontology it contains a small number of direct branches. An ontologist who has lower inheritance richness will be deep or vertical, and this indicates that the ontologist specializes in a specific domain and contains details related to this domain. This resembles what appears in the results of the present work. On the other hand, the ontology that shows a high value indicates that it is general and not specialized within a specific domain with less detail.

```
Algorithm (3) Taxonomy & Non Taxonomy relation for concept
1. Input: Concept    // the concept extract by Algorithm (2)
2. Output: Taxonomy & Non Taxonomy relation.
3. Begin
4.  Read concept.
5.  Extract triple relation for concept from (WordNet, BabelNet).
6.  Construct tree concept hierarchy
7. Compress the tree  concept hierarchy.
8. Extract other non -Taxonomy relation (antonymy, meronymy) for concept from
(WordNet, BabelNet).
9. Build OWL ontology by using the concepts and relations with protégé.
10. END
```

4.2 Ratio of Class Relations

This measure describes the relations ratio between the classes in the ontology. Our results indicate that the ratio of relationships between classes is high, eventually indicating a high correlation in ontology.

4.3 Axiom and Axiom/class Ratio

This represents the main component in ontology and they state what is true in a domain. Our ontology has high value for this measure which indicates that our results are correct (Table 4).

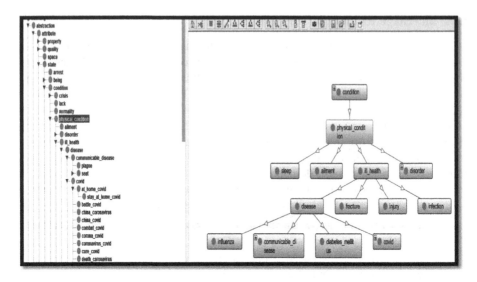

Fig. 3. Protégé API & ontology hierarchy

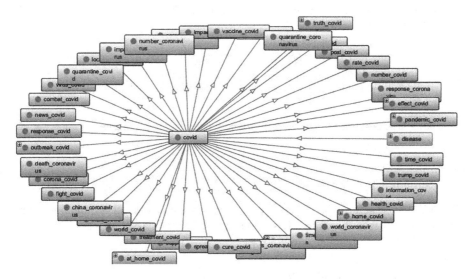

Fig. 4. Covid-19 concept hierarchy

Table 3. Base metrics results for tweets ontology

Base metrics	Value
Axioms	3018
Logical axioms count	941
Class count	950
Total classes count	950
Object property count	2
Total object properties count	2
Data property count	0

Table 4. Schema metrics results for tweets ontology

Schema metrics	Value
Inheritance richness	1.004706
Class/relation ratio	1.903294
Axiom/class ratio	3.550588
Annotation assertion axioms count	1226

5   Conclusion

Due to the different terms used to express the same topic in a Twitter post, it becomes difficult to build applications for processing such as detecting information diffusion, retrieval of information, information extraction, and automatic translation using previous mining methods to find the matching between words or sentences.

The present work solves this problem by generating an ontology for the Twitter post with the hashtag covid-19 using Ontology learning. This is a challenging research field, intersecting the field of machine learning, natural language processing, data and text mining, knowledge representation, and natural language processing.

Due to using combining of machine learning, linguistic and statistical information metrics. The technique showed high efficiency in extracting terms, which extract 900 single terms and 180 multi-word terms. Afterwards, the term input to the disambiguation technique.

The proposed Words Sense Disambiguation method adapted on WordNet lexical to identify the most appropriate senses of the term in the text according to their contexts, the most appropriate sense was chosen as the sense which has the highest cumulative similarity value with senses of all other words in a sentence. The results showed by using two semantic dictionaries Wordnet, BabelNet, the compressed tree technique and Words Sense Disambiguation a high accuracy and depth in extracting the semantic relationships of concepts in a specific domain as covid-19 tweet compared to the previous rules-based methods.

The ontology is built semi-automatically from scratch which contains concepts in the English language that is used for the purposes of matching words and sentences. It is possible in the future to build an ontology that includes more than one language used to match the meaning of two words in different languages and use it in translation or data match based on semantic information.

References

1. World Health Organisationhttps. https://covid19.who.int. Accessed 21 Apr 2021
2. Chew, C., Eysenbach, G.: Pandemics in the age of twitter: content analysis of tweets during the 2009 H1N1 Outbreak. PLoS ONE 5(11), e14118 (2010). https://doi.org/10.1371/journal.pone.0014118
3. Rufai, S., Bunce, C.: World leaders' usage of twitter in response to the Covid-19 pandemic: a content analysis. J. Publ. Health 42(3), 510–516 (2020). https://doi.org/10.1093/pubmed/fdaa049
4. Somodevilla, M.J., Ayala, D.V., Pineda, I.: An overview on ontology learning tasks. Comput. Sist. 22(1), 137–146 (2018). https://doi.org/10.13053/cys-22-1-2790
5. Maricela, B., Luis, F., Hoyos, R.: Methodology for ontology design and construction. Acc. Admin. 64(4), 1–24 (2019)
6. Fatima, N., Al-Aswadi, H.Y.: Automatic ontology construction from text: a review from shallow to deep learning trend. Artif. Intell. Rev. 53, 3901–3928 (2020)
7. Gruber, T.R.: Toward principles for the design of ontologies used for knowledge sharing? Int. J. Human-Comput. Stud. 43(5–6), 907–928 (1995). https://doi.org/10.1006/ijhc.1995.1081
8. Maedche, A., Neumann, G., Staab, S.: Bootstrapping an Ontology-based Information Extraction System. Studies in Fuzziness and Soft Computing, Intelligent exploration of the web, pp. 345–359. Springer (2003)
9. Gómez-Pérez, A., Manzano-Macho, D.: A survey of ontology learning methods and techniques .Deliverable 1.5, IST Project IST-20005–29243- OntoWeb (2003)
10. Shamsfard, M., Barforoush, A.: The state of the art in ontology learning. Knowl. Eng. Rev. 18(4), 293–316 (2003)
11. Xing, J., Ah-Hwee, T.: CRCTOL: a semantic-based domain ontology learning system. J. Am. Soc. Inform. Sci. Technol. (2010). https://doi.org/10.1002/asi.21231
12. Saeed, A., Tarek, H.: Arabic ontology learning from un-structured text IEEE/WIC/ACM International Conference on Web Intelligence (WI) (2016).https://doi.org/10.1109/WI.2016.0082
13. Sourish, D., Ankur, P., Gaurav, M., Priyansh, T., Jens, L.: Formal Ontology Learning from English IS-A Sentences Computer Science. ArXiv, Corpus ID: 3635019 (2018)
14. Stone, B., Dennis, S., Kwantes, P.: Comparing methods for single paragraph similarity analysis. Topics. Cogn. Sci. 3(1), 92–122 (2011)
15. Katerina, F., Sophia, A., Hideki, M.: The C-value/NC-value domain independent method for multi-word term extraction. J. Nat. Lang. Process. 6(3), 145–179 (1999)
16. Princeton University. http://wordnetweb.princeton.edu/perl/webwn. Accessed 15 Jan 2021
17. Sapienza University of Rome. https://babelnet.org/. Accessed 20 Jan 2021
18. Gerhard, B., Miguel, V.: Automatic Topic Hierarchy Generation Using WordNet. Published in DH 2012, Computer Science, Corpus ID: 51875187 (2012)
19. Stanford Junior University. http://protege.stanford.edu. Accessed 20 Mar 2021
20. University of Rostock. https://ontometrics.informatik.uni-rostock.de/ontologymetrics. Accessed 25 Apr 2021

Audio Compression Using Transform Coding with LZW and Double Shift Coding

Zainab J. Ahmed$^{1(\boxtimes)}$ (ID) and Loay E. George^2 (ID)

^{1} Department of Biology Science, College of Science, University of Baghdad, Baghdad, Iraq
^{2} University of Information Technology and Communications, Baghdad, Iraq

Abstract. The need for audio compression is still a vital issue, because of its significance in reducing the data size of one of the most common digital media that is exchanged between distant parties. In this paper, the efficiencies of two audio compression modules were investigated; the first module is based on discrete cosine transform and the second module is based on discrete wavelet transform. The proposed audio compression system consists of the following steps: (1) load digital audio data, (2) transformation (i.e., using bi-orthogonal wavelet or discrete cosine transform) to decompose the audio signal, (3) quantization (depend on the used transform), (4) quantization of the quantized data that separated into two sequence vectors; runs and non-zeroes decomposition to apply the run length to reduce the long-run sequence. Each resulted vector is passed into the entropy encoder technique to implement a compression process. In this paper, two entropy encoders are used; the first one is the lossless compression method LZW and the second one is an advanced version for the traditional shift coding method called the double shift coding method. The proposed system performance is analyzed using distinct audio samples of different sizes and characteristics with various audio signal parameters. The performance of the compression system is evaluated using Peak Signal to Noise Ratio and Compression Ratio. The outcomes of audio samples show that the system is simple, fast and it causes better compression gain. The results show that the DSC encoding time is less than the LZW encoding time.

Keywords: Audio · DCT · DWT · LZW · Shift coding

1 Introduction

Modern operating systems support a native format for audio files because digital audio has become as significant as digital images and videos [1]. The speech compression process converts input speech data stream into a smaller size data stream, by removing inherent redundancy related to speech signals. The compression mechanism reduces the overall program execution time and storage area of the processor [2]. The principle objective within the audio coding process is to demonstrate the audio signal with a minimum range of bits by implementing transparent signal reproduction and producing output audio that can't be recognized from the initial signal [3]. DCT and DWT techniques are usually utilized on the speech signal. The DCT is used for data compression because in this technique the rebuilding of the audio signal can be done very

© Springer Nature Switzerland AG 2021
A. M. Al-Bakry et al. (Eds.): NTICT 2021, CCIS 1511, pp. 17–32, 2021.
https://doi.org/10.1007/978-3-030-93417-0_2

accurately [4], while DWT is very appropriate for speech compression because of the localization feature of wavelet along with the time-frequency resolution property [5]. There are many mechanisms with different amounts of performance that have been applied in the last few years. Drweesh and George [6] employed the biorthogonal tab 9/7 to design a perfect and low complexity audio compression scheme. The suggested system is composed of the audio data normalization, wavelet transform, quantization, adjusted run-length encoding, and high order shift encoder. In decoding processes, the last operation is to decrease the impact of quantization noise through the post-processing filtering process. Usually, this noise is distinguished at the low energetic parts of the audio signal. The outcomes showed that the promising compression system and the CR are expanded with the increment in the number of passes (Npass) in the wavelet. The post-processing step has encouraging results by improving the quality of the recreated audio file signal and enhancement of the fidelity level of the recreated audio signal when PSNR is under 38 Db. Romano et al. [7] used biorthogonal wavelet filters to design a new method for analyzing and compressing speech signals. The thresholding methods are applied to remove some unnecessary details of the signal, then acquiring a lossy compression that makes notably reduces the audio bit-stream length, without the distortion of the quality of the resulting sound. A comparison was made between this innovative compression method with a typical VoIP encoding of the human voice, confirming how using wavelet filters may be suitable, especially in CR without making a significant weakness in audio signal quality for listeners. Salau [8] used DCT with temporal auditory masking (TAM) to perform an algorithm for the compression of audio signals. The outcomes illustrated that the suggested algorithm gives a CR of 4:1 of the original signal. This was assisted by the recording device's negligence of background noise. Ahmed et al. [9] suggested an audio compression system that depends on a combined transform coding system; the bi-orthogonal transform to analyze the input audio signal file into low sub-band and many high sub-bands, the DCT applied to these sub-bands, the progressive hierarchical quantization performed to result of the combined transform stage, then traditional run-length encoding, and lastly, LZW algorithm applied to produce the results. The results show the good performance of this system and the output compression results appear a wonderful reduction in audio file size with good PSNR.

An audio compression problem is to detect an effective, fast and functional method to remove various redundancies between data of the audio file. So, the research work aims to develop a system for compression audio using both Wavelet and DCT transform. The target "better compression with good fidelity" has been taken into consideration during the design phase. The main contributions of this system are: (1) Two entropy encoder is given: LZW and develop version of shift coding called DSC, (2) Applies run-length encoding depend on zeros and runs in audio sequences, (3) Comparison between wavelet and DCT with two entropy encoder, and (4) Comparison in encoding and decoding time between two entropy encoder.

2 The Proposed Methodology

The proposed audio compression system contains many stages; detail of these stages is described in the following subsections (as shown in Fig. 1).

2.1 Audio Data Loading

The audio file is loaded for obtaining the basic file information description signal detailing through viewing the header data. The header data has the sample numbers, channel numbers, and the resolution of sampling. Thereafter in an array, these audio sample data are stored, then the equation [1] is used to normalize to the range [−1, 1]:

$$W_n = \begin{cases} \dfrac{W(i)-127.5}{127.5} & \text{if Sampling resolution} = 8\,\text{bit} \\ \dfrac{W(i)}{32768} & \text{if Sampling resolution} = 16\,\text{bit} \end{cases} \tag{1}$$

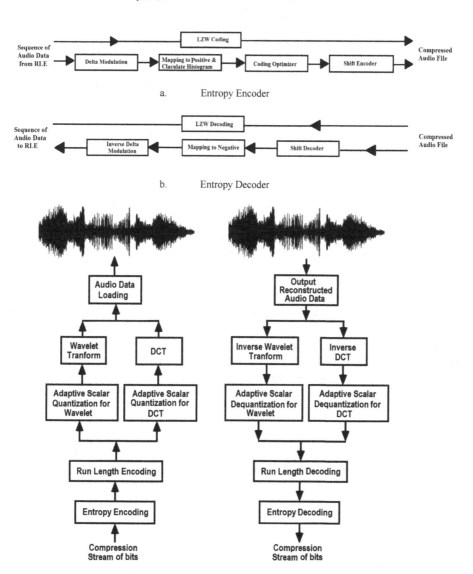

Fig. 1. The layout of the proposed system

2.2   Transform Technique

Transform coding is the process that transforms the signal from its spatial domain to various representations often to the frequency domain [10]. It is used to provide a result with a new set of smaller values data. In this audio compression system, it was implemented one of two ways of transform coding; the bi-orthogonal (tap-9/7) wavelet transform, and the DCT.

Bi-orthogonal Wavelet Transform (Tap 9/7). The bi-orthogonal wavelet transform is performing by applying a lifting scheme after applying the scaling scheme. The lifting step is consisting of a sequence of phases and is characterized into three phases: Split phase, Predict phase, and Update phase [11].

DCT (Discrete Cosine Transform). The signal is partitioned into non-overlapping blocks in DCT, and then each block is handled separately and transformed into AC and DC coefficients in the frequency domain [12]. The following equations applied this process [13]:

$$C(u) = \alpha(u) \sum_{i=0}^{Num} A(i) \cos(\frac{u\pi(2i+1)}{2Num}) \tag{2}$$

$$\text{Where Alpha is, } \alpha(u) = \begin{cases} \sqrt{1/Num} & \text{if } u = 0 \\ \sqrt{2/Num} & \text{if } u \neq 0 \end{cases} \tag{3}$$

Where u = 0.., Num–1 and C (u) appointed to the uth coefficient of the DCT, and A () appointed to a set of Num audio input data values.

2.3   Quantization

It is the process that rounding floating-point numbers to integers and thus in the coding process, fewer bits are necessary to store the transformed coding coefficients [14]. The main goal of this process is to make high-frequency coefficients to zero by reducing most of the less important of these coefficients [15]. Every element in the DCT array is quantized, this process is done by dividing the coefficients of each block by its quantization step (Q_S). The equation [4, 5] is to implement the quantization process:

$$DCT_Q(x, y) = \text{round}\left(\frac{DCT(x, y)}{Q_s}\right) \tag{4}$$

Where the Q_s are used to quantize the coefficients and calculated by the following equations:

$$Q_s = \begin{cases} Q_0 & \text{for DC coefficient} \\ (Q_1 * (1 + \alpha(u,\ v)) & \text{for AC coefficients} \end{cases} \tag{5}$$

The produced wavelet detailed and approximate coefficients in each subband are quantized using progressive scalar quantization where this quantization acts by performing quantization in a hierarchal shape. In the wavelet transform, every pass is quantized with various scales beginning with large quantization scales, and when the number of passes increases then quantization scales decrease drastically. The used equation for uniform quantization is [6]:

$$Wavelet_q(i) = round\left(\frac{Wavelet(i)}{Q}\right) \qquad (6)$$

Where, $Q = Q_{low}$; $Q_{high}(n)$ when the quantization is applied on approximation subband and nth high subband, Wave(i) is the input coefficient of the wavelet transform, $Wavelet_q(i)$ is the corresponding quantization wavelet. For detail subbands, the quantization step is:

$$Q_{high}(n-1) = \beta \times Q_{high}(n) \qquad (7)$$

Where β is the rate of increase of the applied quantization step.

2.4 Run-Length

It is a simple form of lossless compression [16]. It is composed of outputting the lengths of the sequences with successive repetitions of a character in the input and is usually applied to compress binary sequences [17]. The steps for applying the run-length are:

1. Take the non-zero values and put them in a new vector in the sequence called a non-zero sequence.
2. Replace the original location of non-zero values in the input sequence with ones and make the zeros values and ones values in the array and then set it in a new different vector called Binary Sequence.
3. For the series of zeros and ones, run-length encoding is utilized
4. Just the non-zero sequences and their length are encoded and stored.

The following example illustrates the process:

Input Sequence = {3, 7, –3, 0, 1, 2, 0, 0, 0, 0, –1, 2, 4, 1, 0, 0, 0, 0, 0, 0, 1, –2, 0, 0, 0, 0, 0, 0}.
Non-Zero Sequence = {3, 7, –3, 1, 2, –1, 2, 4, 1, 1, –2}.
Binary Sequence = {1, 1, 1, 0, 1, 1, 0, 0, 0, 0, 1, 1, 1, 1, 0, 0, 0, 0, 0, 0, 1, 1, 0, 0, 0, 0, 0, 0}.
Run Length = {1, 3, 1, 2, 4, 4, 6, 2, 6}.
1 = because the start element is not zero.

2.5 Entropy Encoder

The lossless compression is applied as the last step of the proposed audio compression to save the quantized coefficients as the compressed audio. In this paper, one of the two coding schemes has been utilized to compress the audio data; the first one is LZW coding, and the second one is Double Shift Coding (DSC).

Lempel-Ziv-Welch (LZW). LZW is a popular data compression method. The basic steps for this coding method are; firstly the file is read and the code is given to all characters. It will not allocate the new code when the same character is located in the file and utilize the existing code from a dictionary. This operation is continued until the end of the file [18]. In this coding technique, a table of strings is made for the data that is compressed. During compression and decompression, the table is created at the same time as the data is encoded and decoded [19].

Double Shift Coding (DSC). The double shift is an enhanced version for the traditional shift coding (that based on using two code words, the smallest code word is for encoding the most redundant set of small numbers (symbols), and the second codeword, which is usually longer, is for encoding other less redundant symbols). The double shift encoding is used to overcome the performance shortage due to encoding the set of symbols that have histograms with very long tails. The involved steps of shift encoding are dividing the symbols into three sets; the first one consists of a small set of the most redundant symbols, and the other two longs less redundant sets are encoded firstly by long code then they categorized into two sets by one leading shift bit {0,1}. The applied steps for implementing the enhanced double shift coding consist of the following steps (the first two steps are pre-processing steps to adjust the stream of symbols and making it more suitable for double shift coding):

Delta Modulation. The primary task of delta modulation is to remove the local redundancy existent between the inter-samples correlation. The existence of local redundancy leads to the range in the sample values being large and the range will be smaller when it is removed. The difference between every two neighboring pixels is calculating using the following equation:

$$Wav_{DMod}(x) = Wav(x) - Wav(x - 1) \qquad (8)$$

Where x = 1... Size of the data array.

Mapping to Positive and Calculate the Histogram. The mapping process is necessary to remove the negative sign from values and to reduce difficulties in the coding process when storing these numbers. This is a simple process performed by the negative numbers are represented to be positive odd numbers while the positive numbers are represented to be even numbers. The mapping process is implemented by performing the following equation [20]:

$$M_i = \begin{cases} 2M_i & \text{if } M_i \geq 0 \\ -2M_i + 1 & \text{if } M_i < 0 \end{cases} \qquad (9)$$

Finally, the histogram of the output sequence and maximum value in the sequence are calculated.

Coding Optimizer. This step is used to define the length of the optimal three code words used to encode the input stream of symbols. The optimizer uses the histogram of symbols, the search for the optimal bits required to get fewer numbers of bits for encoding the whole stream of symbols. The equation of determining the total numbers of bits are:

$$TotBits = n_1 \sum_{i}^{2^{n1}-2} His(i) + (n_1 + n_2 + 1) \sum_{i=2^{n1}-1}^{i=2^{n1}+2^{n2}-2} His(i)$$

$$+ (n_1 + n_3 + 1) \sum_{i=2^{n1}+2^{n3}-1}^{Max} His(i)$$

(10)

Where, (n_1, n_2, n_3) are integer values bounded the maximum symbol value (Max); such that:

$$n \in \log_2(Max), n_1 \in [2, n-1], \ n_2 \in [2, n], \ Son3 \rightarrow \lceil \log_2(Max - 2^{n1} + 2^{n3} - 1) \rceil$$

(11)

According to the objective function of total bits (equation *), a fast search for the optimal values of (n_1, n_2) is conducted to get the least possible values for total bits.

Shift Encoder. Finally, the double shift encoder is applied as an entropy encoder due to the following reasons:

- This technique is characterized by simplicity and efficiency in both the encoding and decoding stages.
- In the decoding process, the size of its corresponding overhead information that is needed is small.
- It is one of the simple types that belong to Variable Length Coding (VLC) schemes. In the VLC methods, short codewords are assigned to symbols that have a higher probability.
- In this coding technique, the statistical redundancies are exploiting to improve attained compression efficiency.

The compression audio file can be rebuilt to make the original audio signal. This process is done by loading the overhead information and the audio component values which are stored in a binary file and, then, performing in reverse order the same stages that are applied in the encoding stage. Figure 1b explained the main steps of the decoding process.

3 Results and Discussion

A set of eight audio files have been conducted to test, determine and evaluate the suggested system performance. The features of these files are shown in Table 1. The waveform patterns of these audio samples show in Fig. 2. The performance of the suggested audio compression system was studied using several parameters. All audio samples are evaluated by two schemes; the first scheme illustrates the results of compression when the LZW method is applied and the second scheme illustrates the results of compression when the DSC method is applied. The test results are assessed and compared depend on PSNR and CR. Also, the real-time encoding and decoding process was assessed. The CR, MSE (Mean Square Error), and PSNR are defined as follows [21]:

$$CR = \frac{\text{Uncompressed File Size}}{\text{Compressed File Size}} \qquad (12)$$

$$MSE = \frac{1}{S} \sum_{k=0}^{S-1} (i(k) - j(k))^2 \qquad (13)$$

$$PSNR = 10 log_{10} \left[\frac{\text{The Dynamic Range of Audio}^2}{MSE} \right] \qquad (14)$$

Where, $i(k)$ is the value of the input audio data, $j(k)$ is the value of the reconstructed audio file, S represents the number of the audio samples, and the dynamic range of the audio file is 255 when 8-bits sample resolution, and equal to 65535 for 16-bits sample resolution.

The control parameters impacts have been explored to analyze the results of the proposed system:

- N_{Pass} is the pass number in wavelet transform.
- Blk is the block size in DCT.
- Q_0, Q_1, and α are the quantization control parameters.
- The sampling rate is how many times per second a sound is sampled.
- Sampling resolution is the number of bits per sample.

The range of the control parameters appears in Table 2, while the supposed default values of these control parameters appear in Table 3. The default values are specified by making a complete set of tests then selecting the best setup of parameters. Each parameter is tested by changing its value while setting other parameters constant at their default values. Figure 3a displays the effect of N_{pass} on CR and PSNR. The choice N_{pass} equal to 2 is recommended because it leads to the best results. Figures 3b and 4) display the effects of Blk, Q_0, Q_1, and α on PSNR and CR on the system performance for audio test samples, the increase of these parameters leads to an increase in the attained CR while decreasing the fidelity level. Figure 5 displays the effect of sampling rate on PSNR and CR, the outcomes specified that the 44100 sampling rate obtains higher CR while the lower sampling rate cases obtain lower CR. Figure 6 displays the effect of sampling resolution on the performance of the system. The outcomes specified that the sampling resolution has a significant impact and the 16-bit sampling resolution displays the best results. Figure 7 displays the effect of run-length on the coding system; it gives the best result in compression and fidelity. For comparison purposes, Fig. 8 displays the attained relationship (i.e., PSNR versus CR) for the DCT method and the wavelet method. The DCT method outperforms the performance of the wavelet method. Table 4 displays the comparison between LZW and DSC on encoding and decoding time. So, the time of the DSC method is the same or less than the time of the LZW method in the case of encoding or decoding.

Table 1. The characteristics of the audio test samples

Attributes	Audio samples							
	Test (1)	Test (2)	Test (3)	Test (4)	Test (5)	Test (6)	Test (7)	Test (8)
Sampling rate (kHz)	44100	44100	44100	44100	44100	44100	44100	44100
Sample resolution (bps)	8	8	8	8	16	16	16	16
Size (KB)	635 Kb	58 Kb	77 Kb	862 Kb	948 Kb	447 Kb	367 Kb	374 Kb
Audio type	Soft music	Animal (Tiger)	Dialog (Male)	Azan Sound	Quran	chorus Sound	Laugh	Orchestral

Table 2. The range values of the control parameters

Parameter	Range
N_{pass}	[2, 5]
Blk	[20, 200]
Q_0	[0.01, 0.12]
Q_1	[0.01, 0.24]
A	[0.01, 0.2]
Sampling rate	{11025, 22500, 44100}
Sampling resolution	{8,16}

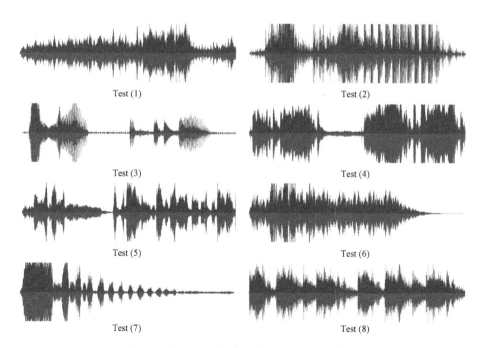

Test (1)

Test (2)

Test (3)

Test (4)

Test (5)

Test (6)

Test (7)

Test (8)

Fig. 2. The waveform of the tested wave files

Table 3. The best PSNR and CR for default values for the suggested control parameters

a.   Audio samples with sample resolution= 8 bps and sample rate=44100 kHz

Wavelet

Audio Samples	PSNR	CR		N_{pass}	Q_0	Q_1	α
		LZW	DSC				
Test (1)	36.759	10.377	9.648				
Test (2)	37.008	6.296	5.873	2	0.09	0.18	0.15
Test (3)	39.169	9.808	8.479				
Test (4)	37.586	9.857	8.068				

DCT

Audio Samples	PSNR	CR		Blk	Q_0	Q_1	α
		LZW	DSC				
Test (1)	40.014	27.922	26.244				
Test (2)	38.747	15.801	13.648	140	0.12	0.12	0.04
Test (3)	39.962	29.029	28.024				
Test (4)	44.876	22.502	21.343				

b.   Audio samples with sample resolution= 16 bps and sample rate=44100 kHz

Wavelet

Audio Samples	PSNR	CR		N_{pass}	Q_0	Q_1	α
		LZW	DSC				
Test (5)	37.132	7.632	6.282				
Test (6)	37.034	7.420	6.653	2	0.08	0.19	0.09
Test (7)	39.340	9.663	8.831				
Test (8)	36.826	6.361	5.214				

DCT

Audio Samples	PSNR	CR		Blk	Q_0	Q_1	α
		LZW	DSC				
Test (5)	39.033	26.332	27.645				
Test (6)	39.946	22.981	21.896	120	0.12	0.16	0.01
Test (7)	40.483	28.944	27.189				
Test (8)	39.906	23.686	22.833				

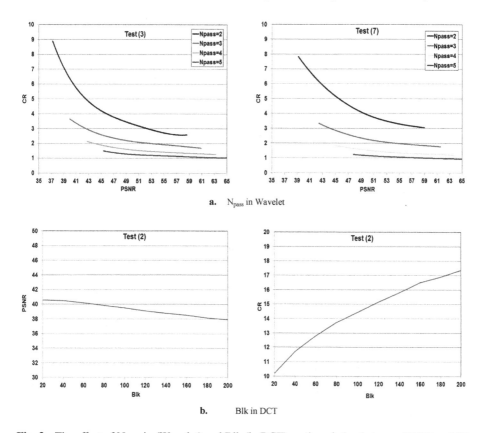

a. N_{pass} in Wavelet

b.   Blk in DCT

Fig. 3. The effect of N_{pass} in (Wavelet) and Blk (in DCT) on the relation between PSNR and CR

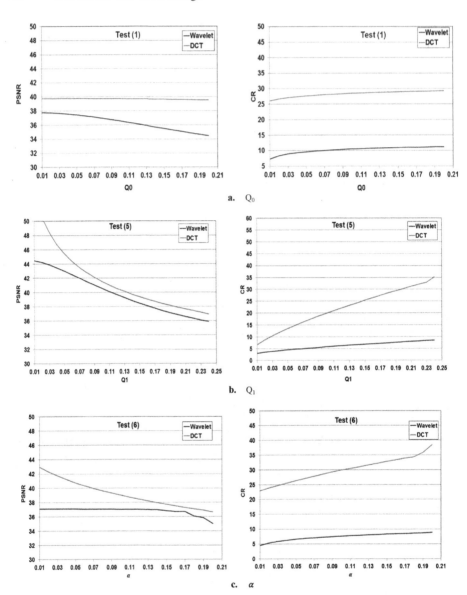

Fig. 4. The effect of (Q_0), (Q_1) and (α), on the relation between PSNR and CR

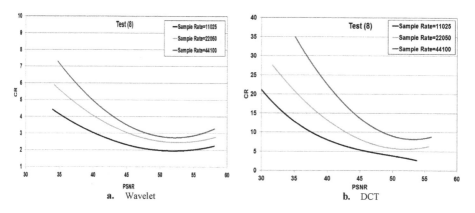

Fig. 5. The effect of sampling rate on the relation between PSNR and CR

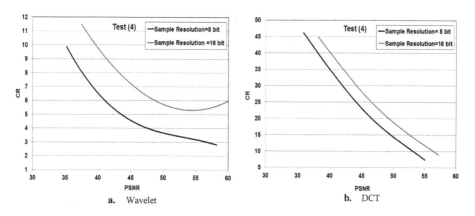

Fig. 6. The effect of sampling resolution on the relation between PSNR and CR

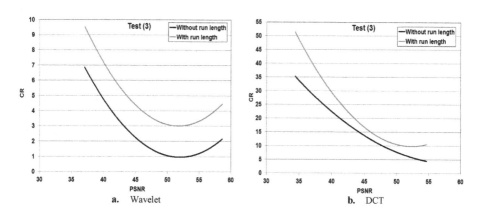

Fig. 7. The effect of run-length on the compression results

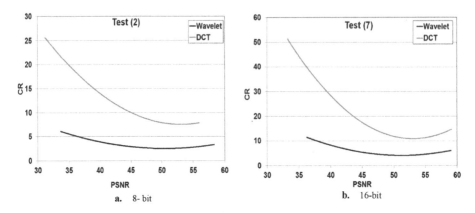

Fig. 8. A comparison between the results of DCT and Wavelet audio compression methods

Table 4. A comparison between LZW and DSC in encoding and decoding time in wavelet and DCT

	Wavelet					DCT			
Audio Samples	LZW		DSC		Audio Samples	LZW		DSC	
	ET	DT	ET	DT		ET	DT	ET	DT
Test (1)	0.689	0.333	0.687	0.155	Test (1)	1.159	0.963	1.051	1.045
Test (2)	0.093	0.056	0.064	0.021	Test (2)	0.125	0.118	0.117	0.117
Test (3)	0.081	0.035	0.062	0.023	Test (3)	0.155	0.152	0.153	0.152
Test (4)	0.961	0.475	0.830	0.494	Test (4)	1.682	0.169	1.511	1.510

a.   8- bit

	Wavelet					DCT			
Audio Samples	LZW		DSC		Audio Samples	LZW		DSC	
	ET	DT	ET	DT		ET	DT	ET	DT
Test (5)	0.647	0.312	0.325	0.136	Test (5)	0.871	0.871	0.849	0.845
Test (6)	0.225	0.156	0.223	0.150	Test (6)	0.408	0.404	0.407	0.402
Test (7)	0.159	0.084	0.119	0.063	Test (7)	0.324	0.311	0.320	0.310
Test (8)	0.242	0.112	0.117	0.115	Test (8)	0.339	0.315	0.299	0.233

b.   16-bit

Table 5. Comparison of the performance results on(test 5) file

Articles	Description	CR	PSNR
[9]	sample resolution = 16 bps and sample rate = 44100 kHz	14.918	40.814
Proposed system		27.645	39.033

4 Comparisons with Previous Studies

It is difficult to make a comparison between the results of sound pressure, due to the lack of the same samples used in those researches, since they are not standardized and each research used samples different from others in other research. Because of that, a comparison was made between the results of the same audio file (test 5) used in a previous search [9], and the results showed in Table 5 a very good CR and PSNR.

5 Conclusion

From the conducted test results on the proposed audio compression system, the compression method achieved good results in terms of CR and PSNR. The increase in the quantization step (i.e.; Q_0, Q_1, and α) and block size (Blk) lead to an increment in the attend CR while a decrease in the PSNR fidelity value. DCT compression is better than bi-orthogonal wavelet transform in terms of CR and PSNR. The use of run-length displays a significant improvement in CR. DSC compression method showed excellent results for CR and very close to the LZW compression method at the same PSNR. When the proposed DSC technique is applied, the encoding and decoding time has been reduced compared to the LZW technique.

As a future suggestion additional enhancements can be made to the proposed entropy encoder system to increase the CR. Also, another coding technique such as Arithmetic Coding can be performed to determine the performance of the proposed encoder system.

References

1. Raut, R., Sawant, R., Madbushi, S.: Cognitive Radio Basic Concepts Mathematical Modeling and Applications, 1st edn. CRC Press, Boca Raton (2020)
2. Gunjal, S.D., Raut, R.D.: Advance source coding techniques for audio/speech signal: a survey. Int. J. Comput. Technol. App. **3**(4), 1335–1342 (2012)
3. Spanias, A., Painter, T., Atti, V.: Audio Signal Processing and Coding. Wiley, USA (2007)
4. Rao, P.S., Krishnaveni, G., Kumar, G.P., Satyanandam, G., Parimala, C., Ramteja, K.: Speech compression using wavelets. Int. J. Mod. Eng. Res. **4**(4), 32–39 (2014)
5. James, J., Thomas, V.J.: Audio compression using DCT and DWT techniques. J. Inf. Eng. App. **4**(4), 119–124 (2014)
6. Drweesh, Z.T., George, L.E.: Audio compression using biorthogonal wavelet, modified run length, high shift encoding. Int. J. Adv. Res. Comput. Sci. Softw. Eng. **4**(8), 63–73 (2014)
7. Romano, N., Scivoletto, A., Połap, D.: A real-time audio compression technique based on fast wavelet filtering and encoding. In: Proceedings of the Federated Conference on Computer Science and Information Systems, IEEE, vol. 8, pp. 497–502, Poland (2016)
8. Salau, A.O., Oluwafemi, I., Faleye, K.F., Jain, S.: Audio compression using a modified discrete cosine transform with temporal auditory masking. In: International Conference on Signal Processing and Communication (ICSC), IEEE, pp. 135–142, India (March 2019)
9. Ahmed, Z.J., George, L.E., Hadi, R.A.: Audio compression using transforms and high order entropy encoding. Int. J. Elect. Comput. Eng. (IJECE) **11**(4), 3459–3469 (2021)

10. Varma, V.Y., Prasad, T.N., Kumar, N.V.P.S.: Image compression methods based on transform coding and fractal coding. Int. J. Eng. Sci. Res. Technol. **6**(10), 481–487 (2017)
11. Kadim, A.K., Babiker, A.: Enhanced data reduction for still images by using hybrid compression technique. Int. J. Sci. Res. **7**(12), 599–606 (2018)
12. Tsai1, S.E., Yang, S.M.: A fast DCT algorithm for watermarking in digital signal processor. Hindawi. Math. Probl. Eng. **2017**, 1–7 (2019)
13. Ahmed, N., Natarajan, T., Rao, K.R.: Discrete cosine transform. IEEE Trans. Comput. **C–23**(1), 90–93 (1974)
14. Tiwari, P.K., Devi, B., Kumar, Y.: Compression of MRT Images Using Daubechies 9/7 and Thresholding Technique. In: International Conference on Computing, Communication and Automation, IEEE, pp. 1060–1066, India (2015)
15. Raid, A.M., Khedr, W.M., El-dosuky, M.A., Ahmed, W.: JPEG image compression using discrete cosine transform - a survey. Int. J. Comput. Sci. Eng. Surv. **5**(2), 39–47 (2014)
16. Hardi, S.M., Angga, B., Lydia, M.S., Jaya, I., Tarigan, J.T.: Comparative analysis run-length encoding algorithm and fibonacci code algorithm on image compression. Journal of Physics: Conference Series. In: 3rd International Conference on Computing and Applied Informatics, 1235, pp.1–6, Indonesia, September 2018
17. Agulhar, C.M., Bonatti, I.S., Peres, P.L.D.: An adaptive run length encoding method for the compression of electrocardiograms. Med. Eng. Phys. **35**(2013), 145–153 (2010)
18. Kavitha, P.: A survey on lossless and Lossy data compression methods. Int. J. Comput. Sci. Eng. Technol. **7**(3), 110–114 (2016)
19. Btoush, M.H., Dawahdeh, Z.E.: A complexity analysis and entropy for different data compression algorithms on text files. J. Comput. Commun. **6**(1), 301–315 (2018)
20. Ahmed, S.D., George, L.E., Dhannoon, B.N.: The use of cubic Bezier interpolation, biorthogonal wavelet and quadtree coding to compress colour images. Br. J. Appl. Sci. Technol. **11**(4), 1–11 (2015)
21. Salomon, D., Motta, G.: Handbook of Data Compression, 5th edn. Springer, London (2010)

The Effect of Speech Enhancement Techniques on the Quality of Noisy Speech Signals

Ahmed H. Y. Al-Noori$^{(\boxtimes)}$ ⓘ, Atheel N. AlKhayyat ⓘ,
and Ahmed A. Al-Hammad ⓘ

Al-Nahrain University, Baghdad Al Jadriya 10072, Iraq
ahmed.alnoori@gmail.com
atheelnowfal964@gmail.com
ahmed.a.al-hammad@nahrainuniv.edu.iq

Abstract. Recently, the human voice has been used widely in different modern applications. However, environmental noise represents one of the significant challenges for these types of speech applications since it affects the quality of the speech signal caused by decreasing the Signal to Noise Ratio (SNR). Speech enhancement techniques represent one of the remedies for this challenge. This paper presents a study of speech cleaning techniques to improve the quality of noisy speech signals in different environmental noise and SNR based on human perception. The study is based on the human auditory system for 50 volunteers. Three types of speech enhancement algorithms are utilised in this study. Despite the results, these enhancement techniques' effects are invariant based on the kind of noise and the value of the Signal to Noise Ratio. Still, the subspace approach performs better than the other two approaches, especially with the high SNR.

Keywords: Speech enhancement · Wiener filter · Spectral subtraction · Subspace approaches · Environmental noise

1 Introduction

One of the most natural types of human-to-human and human-to-machine communication is the human speech signal. It has been utilised in various applications recently, including Automatic Speech Recognition (ASR), Speaker Recognition (voice biometric), speech coding systems, mobile communication, and intelligent virtual assistant. Due to numerous ambient noises, the performance of these speech application systems are severely decarded; hence, the reception task becomes difficult for a direct listener and causes inaccurate transfer of information. Noise suppression and, in turn, enhancement of speech is the main motive of many researchers in the fields of speech signal processing over the decades [1, 2]. Speech enhancement algorithms are designed to increase one or more perceptual characteristics of noisy speech, most notably quality and intelligibility.

In specific applications, the principal objective of speech enhancement algorithms is to increase speech quality while retaining, at the very least, speech intelligibility [2, 3]. Hence, the focus of most speech enhancement algorithms is to improve the quality of speech. Speech enhancement methods seek to make the corrupted noisy speech

© Springer Nature Switzerland AG 2021
A. M. Al-Bakry et al. (Eds.): NTICT 2021, CCIS 1511, pp. 33–48, 2021.
https://doi.org/10.1007/978-3-030-93417-0_3

signal more pleasant to the listener. Furthermore, they are beneficial in other applications such as automatic speech recognition [4, 5].

Improving quality, however, does not always imply that intelligibility would increase. The major cause for this is the distortion imparted on the cleaned speech signal as a result of severe acoustic noise suppression. Speech enhancement algorithms create two forms of distortion: those that impact the speech signal itself (called speech distortion) and those that influence the background noise (called background noise distortion). Listeners appear to be the more impacted by speech distortion when making overall quality judgments of the two types of distortion. Unfortunately, no objective metrics currently exist that correlate high with either distortion or the overall quality of speech enhanced by noise suppression algorithms [6]. Hence, the fundamental challenge in developing practical speech enhancement approaches is to suppress noise without avoiding distortion of speech signal.

Several techniques have been proposed. These techniques can be categorised into three main approaches [7, 8]: Firstly, the spectral subtraction approaches [3, 9–11], which depends on anticipating and updating the spectrum of the noise when there is silence pause in the speech signal, then subtracting the outcome from a noisy speech signal.

Secondly, Statistical model-based techniques, in these techniqes the cleaning speech problems are represented in a statistical prediction framework. These approaches are use a set of measurements, such as the Fourier transform coefficients of the noisy speech signal, to obtain a linear (or nonlinear) estimator of the parameter of interest, referred to as the transform coefficients of the speech signal [7]. Examples of these types of techniques, the Wiener filter [12–14], Minimum mean square error (MMSE) algorithms [15–17], and the maximum-likelihood approach for predicting the spectrum of the clean signal [18–20] and a slew of additional techniques falls under this set. Finally, linear algebra-based algorithms known as Subspace Algorithms: these types are based on linear algebra. The basic notion underlying these algorithms is that the clean signal might be contained within a subspace of the noisy Euclidean Space. Hence, dividing the vector space of a noisy speech signal into a clean signal subspace, which is mostly filled by clean speech, and a noise subspace, which is primarily occupied by noise (Loizou 2013). These algorithms were developed firstly by (Dendrinos et al. 1991)and (Ephraim and Van Trees 1995). This paper aims to study the impact of speech enhancement techniques on improving the quality of speech signal contaminated with different environmental noise based on the auditory system of 50 volunteers. Various Signal to Noise ratios SNRs is used in this investigation. The speech samples (which obtained from SALU-AC speech database) are corrupted with various types of environmental noise (Cafeteria Babble, Construction, and Street Noise).

2 Environmental Noise

Understanding the properties of background noises and the distinctions between noise sources in terms of temporal and spectral characteristics is critical for designing algorithms for suppressing additive noise. The noise signal is known as any unwanted sound signal that you do not need or want to hear. The long-term average spectrum of the five categories of environmental noise is demonstrated in Fig. 1. (Inside the car noise, Cafeteria speech Babble, inside the train trailer, street noise, and white noise signal).

The first kind of monitoring is concerned with a lack of regularity in the spectrum, which gives a unique identity for each type of noise [8]. Noise can be generally classified as stationary noises (also known as Wide Sense Stationary WSS), such as the fan noise coming from PCs, which not change over time.

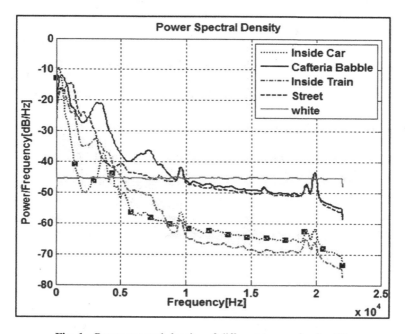

Fig. 1. Power spectral density of different types of noise [8].

Non-stationary noise has spectral characteristics continuously changing over time, such as in cafeteria Babble noise (Fig. 1), making suppression or removing of this type of noise more complicated than suppressing stationary noise. In Cafeteria babble noise, for example, maybe one of the most difficult forms of noise to handle in voice applications since several people chat in the background, which is sometimes mixed with noise from the kitchen. The spectral (and temporal) features of cafeteria noise are continuously changing as customers speak at neighbouring tables and servers engage and converse with them. The Signal to Noise Ratio (also known as speech to noise ratio) (SNR) is defined as the power level of disparity between speech and additive noise. SNR is typically expressed in decibels (dB), such that SNR = 0dB if the speech signal ratio is equal to the additive noise ratio.

In addition, the noise can also be classified into Continuous Noise (engine noise), Intermittent Noise (aircraft flying above your house), Impulsive Noise (explosions or shotgun), and Low-Frequency Noise (air movement machinery including wind turbines). In this paper, three types of non-stationary noise have been used **Cafeteria Speech Babble Noise, Construction Noise, and street noise.**

3   Speech Enhancement Techniques

Speech enhancement techniques, as previously discussed, is concerned with enhancing the perception of the speech signal that has been distorted by ambient noise. In most applications, these techniques' key aim is to increase the quality and intelligibility of the speech signal that is contaminated with environmental noise. In general, The enhancement in quality is more desirable since the technique can decrease listener fatigue, specifically in situations where the listener is exposed to high noise levels for a long time [2]. Since these techniques are applied to reduce or suppress background noise, speech enhancement is also known as the noise suppression algorithms (or speech cleaning) [2]. Various methods for cleaning speech signals and decreasing additive noise levels to increase speech efficiency have been improved in the literature. As stated previously, these strategies can be divided into three categories:

3.1   Spectral Subtraction Approaches

These approaches depend on the consideration that a noisy signal is a combination of both noise and clean speech signals. Consequently, the noise spectrum is calculated during speech pauses. Then the noise spectrum is subtracted from the original signal (noisy signal) to get clean speech [21]. These approaches were first suggested by Weiss et al. [22] and [23]. Consider a noisy signal $y(n)$ which consists of the clean speech $s(n)$ degraded by statistically independent additive noise $d(n)$ as follows:

$$y(n) = s(n) + d(n) \tag{1}$$

It is assumed that additive noise is zero mean and uncorrelated with clean speech. Because the speech signal is non-stationery and time-variant, The speech signal is supposed to be uncorrelated with the background noise. The representation in the Fourier transform domain is given by [24]:

$$Y(\omega) = S(\omega) + D(\omega) \tag{2}$$

The speech can be estimated by subtracting a noise estimate from the received signal.

$$\hat{S}(\omega) = |Y(\omega)| - |\hat{D}(\omega)| e^{j\theta_y(\omega)} \tag{3}$$

Where $|Y(\omega)|$ is the magnitude spectrum, $\theta_y(\omega)$ is the phase (spectrum) of the contaminated noisy signal, $\hat{S}(\omega)$ the estimated clean speech signal.

The estimated speech waveform is recovered in the time domain by inverse Fourier transform $S(\omega)$ using an overlap and add approach. The drawback of this technique is the residual noise.

$$s(n) = IDTFT\{(|Y(\omega)| - |D(\omega)| e^{j\theta(\omega)}\} \tag{4}$$

where $s(n)$ is recovered speech signal.

3.2 Approaches Based on Statistical-Models

These approaches modelled the cleaning speech problem by using a statistical estimating framework. This is based on a set of observations, such as the noisy speech signal's Fourier transform coefficients, to obtain a linear (or nonlinear) estimate of the parameter of interest, known as the transform coefficients of the speech signal [2]. The Wiener filter [25], the maximum likelihood estimator [12], and minimum mean square error (MMSE) algorithms [15] are only a few examples of these sorts of approaches. This paper adopted the Wiener filter as a statistical approach since it represents the most commonly used approach in speech enhancement.

The Wiener filter is one of the most popular noise reduction techniques, and it has been described in a variety of ways and used in various applications. It is based on decreasing the Mean Square Error (MSE) between the estimated signal magnitude spectrum $\hat{S}(\omega)$ and real signal $S(\omega)$. The following is the formula for the best wiener filter [12, 26]:

$$H(\omega) = \frac{S_{s(\omega)}}{S_{s(\omega)} + S_{n(\omega)}} \tag{5}$$

where $S_s(\omega)$ and $S_n(\omega)$ represent the estimated power spectra of the noise-free speech signal and the additive noise, which are assumed uncorrelated and stationary. After measuring the transfer function of the Wiener filter, the speech signal is recovered through [12]:

$$\hat{S}(\omega) = X(\omega) \cdot H(\omega) \tag{6}$$

In a modified form of the Wiener filter, an adjustable parameter α has been used [12].

$$H(\omega) = \left(\frac{S_{s(\omega)}}{S_{s(\omega)} + \beta S_{n(\omega)}}\right)^{\alpha} \tag{7}$$

where β is noise suppression factor.

3.3 Subspace Approaches

These approaches are primarily linear algebra based. The core principle of these methods is that the clean signal may be contained within a subspace of the noisy Euclidean Space. As a result of dividing the vector space of a noisy speech signal into a clean signal subspace, which is mostly filled by the clean speech, and a noise subspace, which is primarily occupied by the noise signal [7, 8]. Then, nullifying the noisy vector variable in the noise subspace to produces the cleaning voice signal. These approaches were suggested by [27, 28]. The signal subspace is plagued by unwanted residual noise. The unwanted noise is supposed to be uncorrelated with the speech signal so that the noisy signal covariance matrix can be written as follows:

$$R_x = R_s + R_w \tag{8}$$

Where R_x is the signal covariance matrix, R_s is the clean speech covariance matrix and R_w is the noise vector with covariance matrix. With these assumptions, the following linear subspace filter is developed to estimate the desired speech vector from the noisy observation:

$$\hat{S} = Hx = Hs + Hw \tag{9}$$

Where Hs and Hw is the filter output and the desired speech after applying filter respectively, the residual error is defined as follows:

$$R = (H - I)s + Hw \tag{10}$$

where r is the residual error. The aim here is to decrease the signal distortion subject to keeping every spectral component of the residual noise in the signal subspace as little as possible.

4 Experimental Setup

Based on the perception of the human auditory system, this paper investigates the impact of speech enhancement approaches for improving the quality of noisy speech signals under varied ambient noise and varying SNR. As previously stated, three types of enhancement are adopted in this paper. The speech signals are adopted in this study were corrupted by three kinds of noise (Cafeteria babble noise, Construction noise, and Street noise) at SNRs (15 dB, 10 dB, and 0 dB). The processed speech signal was presented to regular hearing listeners to evaluate its quality. The results investigated the effect of these filters are invariant based on the type of noise and the value of the signal to noise ratio. Figure 2 show the block diagram of the methodology of this study:

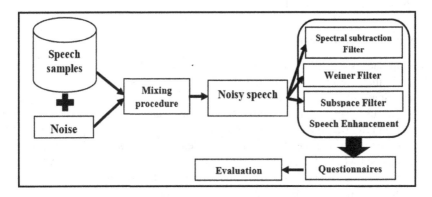

Fig. 2. The block diagram of the suggested study.

The experimental setup of this study can be summarised as follows:

1. Providing speech samples contaminated with different environmental noise and controlled SNR using the mixing procedure. This procedure is described in the next section.
2. Applying Speech Enhancement algorithms on noisy speech sample to produce the filtered speech samples.
3. Evaluating the performance of each speech enchantment algorithms based on the perception of human ears to filtered speech signals. Noisy speech samples and mixture procedure.

4.1 Speech Samples

The experiments were conducted on two speech samples' sets. These speech samples are collected from the University of Salford Anechoic Chamber Database (SALU-AC database) (Fig. 3). One of the database's most distinguishing characteristics is that it includes English speech sample spoken by native and non-native speakers and the recording environment that collected on it since data was gathered in the Anechoic Chamber. The principal purpose of this database was to offer clean speech samples, which make them more efficient while dealing with one adverse condition (such as noise) in isolation from other adverse conditions [29].

Fig. 3. Anechoic Chamber at University of Salford [8].

4.2 Noise Samples

As previously indicated, the speech datasets were generally collected in quiet environments (Anechoic Chamber) with no ambient noise influencing the organised speech signals. Consequently, the noisy speech samples were created by combining speech samples with the aforementioned sources of noise, each with a distinct regulated signal to noise ratio (SNR) (15, 10, and 0 dB).

The following is a summary of the mixing technique[8]:

1. To match the duration of target speech utterances, the noise signal was shortened. The main goal of this phase was to ensure that noise was evenly distributed across the speech signal.

2. Controlling the ratio at which the speech signal and noise were combined by specifying the SNR (in dB). As previously stated, 15dB, 10dB, and 0dB were chosen as mixing ratios because SNR 15 dB is near to clean (i.e., the ratio of speech is high relative to the noise, which is too low) and SNR 0 dB is hardly recognised by the human ear.

3. Normalising the speech and noise signals (this normalisation was done by using the root mean square RMS).

4. Finally, before mixed with the speech signal, the noise signal was scaled to achieve the appropriate SNR. Figure 4 shows a male voice signal that has been mixed with noise at various SNRs.

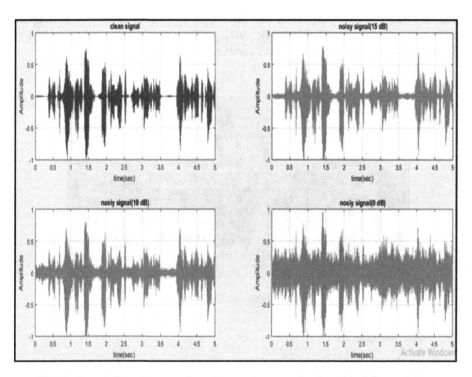

Fig. 4. Speech Sample contaminated by environmental noise with different SNRs.

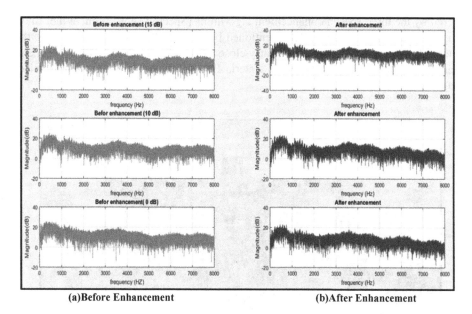

(a)Before Enhancement   (b)After Enhancement

Fig. 5. The wiener filter for speech contaminated with babble noise (a) before enhancement (b) after enhancement with different SNRs.

4.3   Applied Speech Enhancement Algorithms

As mentioned earlier, three speech enhancement algorithms are adopted in this work: Spectral subtraction, Wiener filter, and Sub-space algorithms. Each of these filters is applied to two speech signals. These signals consist of one male speech signal and one female speech signal in order to study the effect of these algorithms of both gender signals. Each speech signal is mixing with different SNRs (15, 10, and 0 dB) for a specific type of noise that discussed earlier. Therefore, in total, we have 18 filtered speech samples, six filtered samples for each enchantment algorithms. Figure 5(a), (b) shows the spectrum of enhanced signals for male signal contaminated with cafeteria babble filtered by Wiener filter algorithm before and after the filtering process.

5   Questionnaires and Evaluations

The last level in this work is to evaluate each enhancement algorithm's performance for each speech signal contaminated with environmental noise mentioned before with each SNR based on the perception of the human auditory system. Fifty volunteers have been chosen (25 males and 25 females) for this purpose. The volunteer's ages are between 20–40 years old. Each one has been checked that has not any hearing issues. The evaluation is conducted in the Multimedia Laboratory at the College of Engineering in Al-Nahrain University, as seen in Fig. 6.

First, each listener has been instructed to listen to the clean signal without any additive noise (original signal). Then, he/she listen to the three filtered speech signals filtered by the three speech enhancement algorithms (Spectral subtraction, Winer filter, and Subspace filter) for each SNR mentioned before. Finally, the volunteer chose the suitable filtered signal that think it clearly close to the original signal. The experiment, then, repeat for each type of noise (Cafeteria Babble, Street, and construction noise) and for each male and female signal.

Fig. 6. Multimedia lab for listening the enhanced speech signal.

5.1   Experimental Results

As mentioned before, this work evaluates the impact of different type of speech enhancement algorithms on speech signals contaminated with different environmental noises and with various SNR depends on the perception hearing of different volunteers. Figure 7 illustrates the bar chart of the evaluation percentages for the three-speech enhancement algorithms in the case of the male speech signal contaminated with cafeteria babble noise. The x-axis represents SNR in 15 dB, 10 dB and 0 dB, respectively, while the y-axis represents the percentage of the evaluation for each filter. Obviously, the subspace filter in the 15 dB and 10 dB have the highest impact (with 37.5% and 54.10% respectively) if compared with the effects of the other two filters (Wiener, and Spectral subtraction filters). In contrary, the Spectral subtraction filter shows the highest impact on the contaminated signal at 0 dB compared with the two other filters with 54.10%. In contrast, the subspace filter are degraded to 12.5% only at the same SNR.

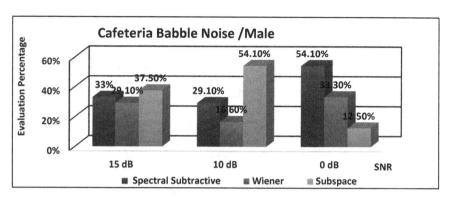

Fig. 7. Bar chart for the effectiveness of speech enhancement Algorithms on male speech signal contaminated with Cafeteria Babble noise.

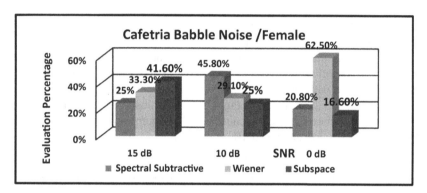

Fig. 8. Bar chart for the effectiveness of speech enhancement Algorithms on female speech signal contaminated with Cafeteria Babble noise.

On the other hand, and as seen in Fig. 8, which represents the effect of the same filters on female speech signal contaminated with Cafeteria babble noise, the sub-space filter still has the higher impact among the other filters at 15 dB with 41.6%. But, at 10 dB, we noticed that the spectral subtraction takes the large effect among filters with 45.8%, while the Wiener filter takes the best evaluation at 0 dB with 62.5%.

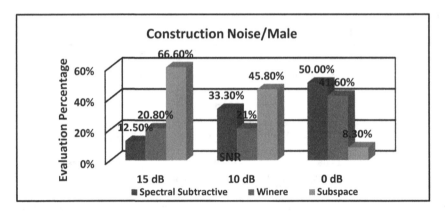

Fig. 9. Bar chart for the effectiveness of speech enhancement Algorithms on male speech signal contaminated with Construction noise.

Figure 9 illustrates the evaluation of the three filters when applied to male speech signal that contaminated with construction noise. It is also clear that the subspace filter still has the best evaluation at 15 and 10 dB compared with 66.6% and 45.88, respectively. On the other hand, Spectral subtraction and wiener filters get the better evaluation compared with sub-space at 0dB with 50% and 41.6%, respectively.

We can notice the same thing in Fig. 10, which illustrates the evaluation of the three filters on the female signal contaminated with the same noise. Subspace algorithm has the highest evaluation at 15 and 10 dB with 50%, 41.6%, respectively.

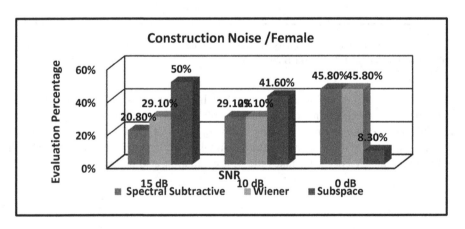

Fig. 10. Bar chart for the effectiveness of speech enhancement Algorithms on female speech signal contaminated with Construction noise.

On the contrary, Spectral subtraction and Wiener filter have the same evaluation at 0dB with 45.8%, while sup-space get only 8.3% at the same SNR. Figure 11 represents the bar chart of the evaluation of the three filters when applied on the male signal when it is contaminated with Street noise. It is clear that the Sub-space filter has the highest effectiveness for cleaning the signal at 15dB with 45.8%. On the other hand, the Wiener filter and Spectral subtraction show the highest effect at 10 dB and 0 dB, respectively.

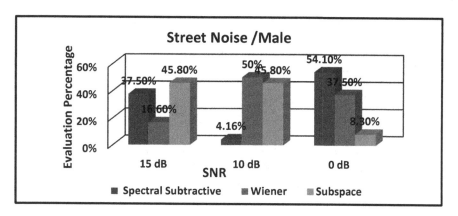

Fig. 11. Bar chart for the effectiveness of speech enhancement Algorithms on male speech signal contaminated with Street noise.

Almost the same effect is noticed on the female signal that contaminated with the same environmental noise (street noise), as demonstrated in Fig. 12. Table 1 demonstrates the overall evaluation for the three filters to improve the quality of noisy signals with different environments. In this table, for both male and female, the Spectral subtractive gave virtually the best results in 0 dB where the noise level is high. In contrast, the Subspace stratifies the best result 15 dB for both male and female.

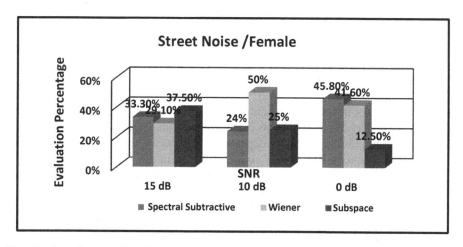

Fig. 12. Bar chart for the effectiveness of speech enhancement Algorithms on female speech signal contaminated with Street noise.

Table 1. Overall evaluation of questionnaires.

Filter	Babble noise					
	Male			Female		
	15 dB	10 dB	0 dB	15 dB	10 dB	0 dB
Spectral	33%	29.1%	54.1%	25%	45.8%	20.8%
Wiener	10%	16.6%	33.3%	33.3%	29.1%	62.5%
Subspace	37.5%	54.1%	12.5%	41.6%	25%	16.6%
Construction noise						
Spectral	12.5%	33.3%	50%	20.8%	29.1%	45.8%
Wiener	20.8%	21%	41.6%	29.1%	29.1%	45.8%
Subspace	66.6%	45.8%	8.3%	50%	41.6%	8.3%
Street noise						
Spectral	37.5%	4.16%	54.1%	33.3%	24%	45.8%
Wiener	16.6%	50%	37.5%	29.1%	50%	41.6%
Subspace	45.8%	45.8%	8.3%	37.5%	25%	12.5%

In summary, and based on human perception human auditory system, we can conclude the following:

1. Each noise has a different effect on the speech signal, making it challenging to select the best speech enhancement approach to improve speech signal quality. However, the Subspace filter shows the best quality improvement among the other filters at 15 dB and 10 dB. On the contrary, Spectral subtraction shows the best improvement for quality at 0 dB.
2. The impact of a speech enhancement algorithm for improving speech signal quality may vary from one signal to another at the same environmental noise at the same SNR. Mainly when the speech signals belong to different signal, as seen in the effect of Subspace filter on the signals contaminated with cafeteria babble noise at 10 dB.

6   Discussion and Conclusion

The essential goal of this work is to study the effects of speech cleaning algorithms for improving the quality of speech signals contaminated with different ambient noise with different signal to noise ratios SNRs depends on the human hear perception for these enhanced speech signals. The speech signals are used in this study are cleaned from various effects of the environment except for the environmental noise and collected from a different gender. Furthermore, the evaluation of the performance of these algorithms is achieved in a professional environment at the Multimedia lab. Three different types of noise are used in this experiment with controlled SNR. The results demonstrate that the Subspace algorithm performs better than the other two filters in terms of enhancing speech quality (Wiener, and Spectral subtraction) in most cases of 15 dB and 10 dB for different types of noise. The main reason that makes the Subspace approach has the higher quality enhancement among the other two approaches is returned to it is natural, which is based on the linear algebra and the way with dealing with environmental noise. However, at 0 dB, the spectral subtraction algorithm shows the best performance for improving speech quality. Furthermore, the effect of these algorithms may vary according to the type of noise and type of speech signal that belongs to males and females. However, this study focuses mainly on a study the quality of cleaning speech but not the intelligibility. What is now needed is a study involving improving adaptive approaches to deal with quality and intelligibility at the same time.

References

1. Upadhyay, N., Karmakar, A.: The spectral subtractive-type algorithms for enhancing speech in noisy environments. In: 2012 1st International Conference on Recent Advances in Information Technology (RAIT) (2012)
2. Loizou, P.C.: Speech Enhancement: Theory and Practice. CRC Press, Boca Raton (2013)
3. Kamath, S., Loizou, P.: A multi-band spectral subtraction method for enhancing speech corrupted by colored noise. In: 2002 IEEE International Conference on Acoustics, Speech, and Signal Processing (ICASSP) (2002)
4. Zhang, X., Wang, Z., Wang, D.: A speech enhancement algorithm by iterating single- and multi-microphone processing and its application to robust ASR. In: 2017 IEEE International Conference on Acoustics, Speech and Signal Processing (ICASSP) (2017)
5. Wang, Z.Q., Wang, P., Wang, D.: Complex spectral mapping for single- and multi-channel speech enhancement and robust ASR. IEEE/ACM Trans. Audio Speech Lang. Process. **28**, 1778–1787 (2020)
6. Soni, M.H., Patil, H.A.: Non-intrusive quality assessment of noise-suppressed speech using unsupervised deep features. Speech Commun. **130**, 27–44 (2021)
7. Loizou, P.C.: Speech enhancement based on perceptually motivated Bayesian estimators of the magnitude spectrum. Speech Audio Process. IEEE Trans. **13**(5), 857–869 (2005)
8. Al-Noori, A.: Robust speaker recognition in presence of non-trivial environmental noise (toward greater biometric security), in School of Computing, Science and Engineering. University of Salford (2017)
9. Weiss, M.R., Aschkenasy, E., Parsons, T.W.: Study and development of the INTEL technique for improving speech intelligibility. NICOLET Scientific Corp Northvale, NJ (1975)
10. Boll, S.: Suppression of acoustic noise in speech using spectral subtraction. IEEE Trans. Acoust. Speech Signal Process. **27**(2), 113–120 (1979)
11. Oh, I., Lee, I.: Speech enhancement system using the multi-band coherence function and spectral subtraction method (2019)
12. Lim, J.S., Oppenheim, A.V.: Enhancement and bandwidth compression of noisy speech. Proc. IEEE **67**(12), 1586–1604 (1979)
13. de la Hucha Arce, F., et al.: Adaptive quantization for multichannel Wiener filter-based speech enhancement in wireless acoustic sensor networks. Wirel. Commun. Mob. Comput. **2017**, 1-15 (2017)
14. Yanlei, Z., Shifeng, O., Ying, G.: Improved Wiener filter algorithm for speech enhancement. Autom. Control. Intell. Syst. **7**(3), 92 (2019)

15. Ephraim, Y., Malah, D.: Speech enhancement using a minimum-mean square error short-time spectral amplitude estimator. IEEE Trans. Acoust. Speech Signal Process. **32**(6), 1109–1121 (1984)
16. Mahmmod, B.M., et al.: Low-distortion MMSE speech enhancement estimator based on Laplacian prior. IEEE Access **5**, 9866–9881 (2017)
17. Wang, Y., Brookes, M.: Speech enhancement using an MMSE spectral amplitude estimator based on a modulation domain Kalman filter with a Gamma prior. In: 2016 IEEE International Conference on Acoustics, Speech and Signal Processing (ICASSP) (2016)
18. McAulay, R., Malpass, M.: Speech enhancement using a soft-decision noise suppression filter. IEEE Trans. Acoust. Speech Signal Process. **28**(2), 137–145 (1980)
19. Kjems, U., Jensen, J.: Maximum likelihood based noise covariance matrix estimation for multi-microphone speech enhancement. In: 2012 Proceedings of the 20th European Signal Processing Conference (EUSIPCO) (2012)
20. Takuya, Y., et al.: Maximum likelihood approach to speech enhancement for noisy reverberant signals. In: 2008 IEEE International Conference on Acoustics, Speech and Signal Processing (2008)
21. Aghazadeh, F., Tahan, A., Thomas, M.: Tool condition monitoring using spectral subtraction and convolutional neural networks in milling process. Int. J. Adv. Manuf. Technol. **98**(9–12), 3217–3227 (2018). https://doi.org/10.1007/s00170-018-2420-0
22. Weiss, M., Aschkenasy, E., Parsons, T.: Study and development of the INTEL technique for improving speech intelligibility. DTIC Document (1975)
23. Boll, S.: Suppression of acoustic noise in speech using spectral subtraction. Acoust. Speech Signal Process. IEEE Trans. **27**(2), 113–120 (1979)
24. Upadhyay, N., Karmakar, A.: Speech enhancement using spectral subtraction-type algorithms: a comparison and simulation study. Proc. Comput. Sci. **54**, 574–584 (2015)
25. Jae, L., Oppenheim, A.: All-pole modeling of degraded speech. IEEE Trans. Acoust. Speech Signal Process. **26**(3), 197–210 (1978)
26. Abd El-Fattah, M.A., et al.: Speech enhancement with an adaptive Wiener filter. Int. J. Speech Technol. **17**(1), 53–64 (2013). https://doi.org/10.1007/s10772-013-9205-5
27. Dendrinos, M., Bakamidis, S., Carayannis, G.: Speech enhancement from noise: a regenerative approach. Speech Commun. **10**(1), 45–57 (1991)
28. Ephraim, Y., Trees, H.L.V.: A signal subspace approach for speech enhancement. IEEE Trans. Speech Audio Process. **3**(4), 251–266 (1995)
29. Al-Noori, A.: Robust speaker recognition in presence of non-trivial environmental noise (toward greater biometric security). University of Salford (2017)

Image Steganography Based on DNA Encoding and Bayer Pattern

Elaf Ali Abbood[(✉)], Rusul Mohammed Neamah,
and Qunoot Mustafa Ismail

Computer Department, Science College for Women, University of Babylon,
Babylon, Iraq
wsci.elaf.ali@uobabylon.edu.iq

Abstract. Steganography is a method of concealing sensitive data while transmitting it over public networks and it is regarded as one of the invisible security techniques. This paper proposed a new method to encrypt and hide a 24-bit color secret image within a larger 24-bit color cover image using the Bayer pattern principle and LSB algorithm. The secret image goes through a series of special steps of DNA encryption. Then hides to ensure more confidentiality and be more difficult for the attackers' attempts. In this method, each pixel in the secret image is hidden in a 3×3 block from the cover image except the center of the block that is used as a public key. These blocks randomly select from the rows and columns of the cover images. The secret image pixels and the center selected block pixels are encrypted in a special method using DNA rules. Then, encoding the resulted encrypted image and encrypted center blocks selected pixels using the DNA-XOR operation. Finally, hiding the resulted image pixels using the Least Significant Bit (LSB) algorithm with locations determined by a modified Bayer pattern. The experimental results show the efficiency of the proposed method to produce a high-quality stego image and the results are acceptable when applying the Structural Similarity Index Metric (SSIM) and Peak Signal to Noise Ratio (PSNR) quality metrics and compared with other method.

Keywords: Information hiding · DNA encoding · Bayer pattern · PSNR · SSIM

1 Introduction

Hiding data is typically related to well-publicized infamous attempts, like secret arranging and coordination of crimes through messages hidden in pictures spread on the general websites [1, 2]. In addition to the many misuse cases, steganography is also utilized in the application of positive operations. For instance, concealed pictures utilized by way of watermarks involve copyright and copyright datum without outwardly disfiguring the picture [3]. In the development of information and communication technology fields, the transmission of confidential information through general channels and websites has become one of the main challenges of our time. Where there are three techniques for concealing information: encryption, concealment, and watermark.

© Springer Nature Switzerland AG 2021
A. M. Al-Bakry et al. (Eds.): NTICT 2021, CCIS 1511, pp. 49–62, 2021.
https://doi.org/10.1007/978-3-030-93417-0_4

The first technique depends on masking the presence of a message. The second technique hides the data as a media format like image, audio, video, and even text in which unauthorized people do not observe the presence of the data in the aforementioned form. The last technology protects copyrights and authorship. Lately, much attention has been given to the science of concealing information [4].

Steganography is a science that aims to include private and confidential data in the cover medium. This process protects copyrights or identifies the identity and securing the data so that the hackers cannot discover the information [5, 6]. Image files are especially popular for medium cover because there is a great deal of redundant space suitable for hiding confidential information. Steganography techniques, by and large, utilize a comparable key to embed confidential messages and create yield information known as stego-image ought to imperceptible of the carrier media to darken the correspondence between the different parties [7].

In this approach, image properties such as a histogram, the SSIM must stay unchanged after the confidential information is covered in it [8]. Encryption is a technique that saves confidential data by encrypting it, so that only an authorized person can read it [9].

The least significant bit (LSB) is the most widely utilized and common image masking technique [10]. Accordingly, the value of the pixel is expanded or reduced by one. If the value of the pixel of the image is changed by "1", it will not alter the appearance of the image. This makes it easier to hide the information, especially when the cover image is larger than the image of the secret message [11].

The Bayer pattern is one of the techniques that used in the proposed method. The Bayer pattern is expressed by a color filter array (CFA). It is considered as an interpolation technique used to find the missed colors in the sensors of the images in digital cameras. As illustrated in Fig. 1, this pattern contains red and blue pixels, which surround green pixels vertically and horizontally. And because the human eye is more sensitive to green, that's why Bayer allocates the largest number of green pixels versus red and blue to produce a better color image [12].

To strength the hiding technique, it has merged with one of the encryptions techniques. DNA encryption is one of the image encryption techniques. DNA encryption uses the DNA molecules to encodes image pixels in a special way that gives big parallelism with little consumption for energy and high storage compactness [13, 14]. Consequently, DNA image encryption methods are considered as one of the straight encryptions methods against the traditional cryptanalysis algorithms [15].

In this work, the secret image is encrypted using a new DNA encryption method. The encrypted color image is hidden within random blocks of the cover image. Every three bands from the secret image pixel are hidden in a block from the cover image. The proposed method depends on the concept of the Bayer pattern in the hiding process after introducing some modifications.

The contributions of this work contain:

1. Increasing the security robustness by encrypting the secret image using a new DNA encryption method. This encoding process increases the safety of the method and it makes the method difficult for hackers to discover confidential information.

2. Hiding each pixel in the secret image in a random 3×3 block using a new method consist using the Bayer pattern as an index map to find the location of the hiding process that adapts the LSB technique [16].
3. Introducing a new hiding information method combined with a new encryption method that efficient to the color images and robust against many types of attacks.
4. The method uses the centers of the blocks selected from the cover image as a public key that uses to select the DNA rule for encoding the secret image and the centers of the blocks. Also, the initial value of generating random blocks within the cover image is considered a secret key. The recipient to determine where the bits of the secret message are hidden in the stego image and to decrypt the encrypted pixels uses these keys.

2   Related Works

In this section, the discerption of recent researches introduced methods of steganography and information hiding techniques. Relying on the least significant bit and chaotic bit algorithm, Hidayet Ogras [11] proposed a system for concealing a gray image inside a larger one, generating a stream of messy bits and XOR using bitwise.

Shumeet Baluja [17] proposed a system for concealing color image into another color image of the same size while preserving the quality of two images as much as possible, using deep neural network training technique simultaneously to create masking and detection processes, and the system was trained on images taken randomly from the imageNet database.

Deepesh Rawat and Vijaya Bhandari [18] proposed a method for concealing the confidential image data into a 24-bit color image using enhanced LSB technology, the technology relies on MSB to embed the confidential image into the LSB of the cover image. In the state of a 24-bit color image, two approaches are described. In the first scheme, the last two Least Significant Bits (LSB) of each level (red, green, and blue) of the carrier image are replaced by the two Most Significant Bit (MSB) of the confidential image. In the second scheme, the last Least Significant Bit (LSB) of each red level is replaced by the first Most Significant Bit (MSB) of the confidential image, and the last two Least Significant Bit (LSB) of each green level by the next two Most Significant Bit (MSB) of the confidential image, then the last three Least Significant Bit (LSB) of the blue level are replaced by the next three Most Significant Bit (MSB) of the confidential image.

Dogan [19] proposed a data-concealing method based on the pixel pairs, and utilize a messy map such as PRNG and specify its output if the addition and subtraction operations are applied to the pixel pairs to mask the data, and the method achieved high load capacity, good uptime, and safety.

Sun [20] proposed a method for concealing confidential data into the carrier image based on optimization of a logistic map and DNA sequence, whereby 2 bits of the confidential message are concealed to edge pixels using the canny algorithm in determining the edges of the cover image.

Khan Muhammad et.al [21] introduced a method of hiding the confidential message in RGB color images using the Hue-Saturation-Intensity (HSI) based on the Least Significant Bit (LSB), where the method converts the image from RGB to Hue-Saturation-Intensity (HSI) and includes the confidential message into it and then returns it to RGB.

S.M. Hardi, et al. [22] proposed a way to hide a secret message after encoding it using the RSA RPrime algorithm to overcome weaknesses in the Least Significant Bit (LSB) algorithm and then embed the encoded message within the color cover image using the Least Significant Bit (LSB) algorithm.

R. Rejani et al. [23] proposed an approach to hiding the secret message after encoding it with one of the encryption algorithms, whereby it uses the RGB mod bit method to see if each character of the message can be embedded into the image and record location into a field of the image metadata itself.

Xinghong Qin, et al. [24] introduced a new masking scheme known as ITE-SYN (based on ITEratively adversarial perturbations onto an SYNchronized directions sub-image), where the method relies on using the current function of concealment the expense of initial costs and then analysing the cover image into a set of non-overlapping sub-images and after including each image, it will be modified costs by following the trend profile adjust the collection and then following sub-image will be included with adjusted costs where all confidential data is included.

Rusul Mohammed Neamah, et al. [25] proposed a way to hide the confidential data after encrypting it using an encryption key and the Xnor gateway after that analyse each pixel in the color carrier image to three channels (red, green, and blue) and specify one of these channels to hide the encoded bit of the secret message using the modified Least Significant Bit (LSB) algorithm.

3   The Proposed Method

In this paper, a newly proposed method is presented for hiding a small 24-bit color image inside a larger 24-bit color image using a developed Bayer pattern. Bayer pattern, as introduced in Sect. 1, is considered as a method of color band interpolation that is used to find the lost band colors during image capturing. The idea of hiding the three secret bands in a random block from a cover image is motivated from the structure of the Bayer pattern. The original and developed Bayer patterns are illustrated in Fig. 1.

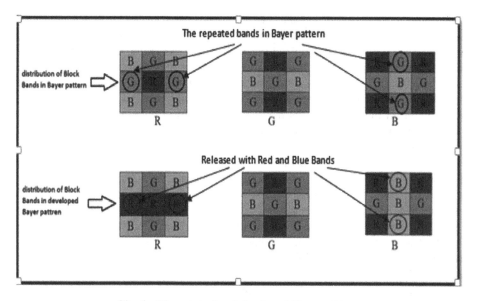

Fig. 1. The original and developed Bayer patterns

If we split the bands of the cover block pixels according to the Bayer pattern, we will find the green band appears in the pattern with the red center and the pattern of the blue center. This case will cause a collision and replication in bands in more than one pattern. For that, a developed Bayer pattern is proposed to prevent the collision and avoid the repeated bands in the patterns. The block diagram of the main steps of the proposed method is illustrated in Fig. 2 and the details of each step are illustrated in Figs. 3 and 4.

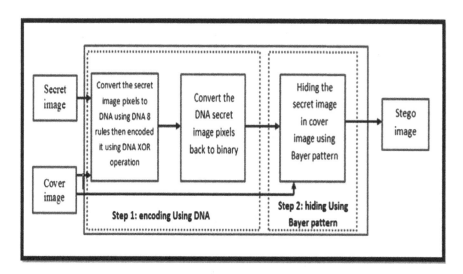

Fig. 2. The main steps of the proposed method

The method is based on breaking each pixel of the secret color image into three bands and each pixel is hidden in a random 3 × 3 block of the colored cover image. The blocks are previously determined in a random way where the number of randomly selected non-overlapped blocks is equal to the secret image pixels. To increase the security of the method, the three bands of each pixel in the secret image will be encoded according to the DNA rules illustrated in Table 1 [25] as in Fig. 3. The least three significant bits of each pixel band (KR, KG, KB) of the block center represents the DNA rule number that used to encode the secret pixel bands (R, G, B) in the secret image respectively. After that, the centers of the randomly selected blocks of the cover image are encoded in the same rules of encoding the secret pixels. Then, the DNA sequence of the secret pixel bands is aggregate with the DNA sequence of the center blocks cover pixel bands using the DNA-XOR operation shown in Table 2 [25]. The resulted DNA-XOR sequence is then converted to binary code again and then to pixels values. As an example of DNA-XOR operation, if the secret DNA encoded sequence is (ATTC) and the center blocks pixels sequence is (GACA) the resulted from encoded sequence using DNA-XOR will be (GTGC). In the extraction procedure, to extract the secret DNA encoded sequence, we apply DNA-XOR operation between the center blocks pixels sequence (GACA) with the resulted encoded sequence (GTGC) and resulted (ATTC) secret DNA sequence.

Table 1. The encoding of DNA rules [25].

	0	1	2	3	4	5	6	7
A	00	00	01	01	10	10	11	11
T	11	11	10	10	01	01	00	00
G	01	10	00	11	00	11	01	10
C	10	01	11	00	11	00	10	01

Table 2. DNA-XOR operation [25]

XOR	A	C	G	T
A	A	C	G	T
C	C	A	C	G
G	G	T	A	C
T	T	G	T	A

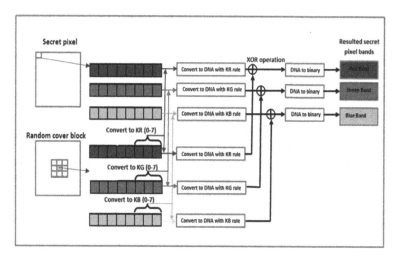

Fig. 3. The DNA encoding step

The bands of each block of a cover image are split relies on developed Bayer patterns to obtain three different blocks of bands as in Fig. 1. The secret pixel bands will be hidden in the block pattern whose center is similar to the secret band as in Fig. 4. Eight bits of each secret band are hidden in the eight bands adjacent to the center band of a block using the Least Significant Bit (LSB) method. It is worth noting that the cover block center is not used to hiding information because it is used as a public key to encrypt the secret pixel with the DNA encoding method.

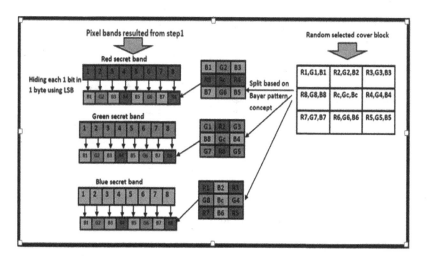

Fig. 4. Hiding step using Bayer pattern and LSB method

The steps of the proposed method for embedding and extraction process are shown below:

A: Embedding procedure:

1. Read cover image and secret image.
2. Find all the blocks in the cover image that randomly selected from the rows and columns of the cover image without overlapping. Then, store their blocks centers in a two-dimensional array as an image.
3. Depending on the concept of the Bayer pattern, the blocks are divided into three blocks as in Fig. 1 to hide each band of the secret pixel in the block whose center band color is identical to the secret band color.
4. Dividing each pixel of the secret image into three bands and converting each one to the DNA code according to Table 1. This encoding rule is determined from the three least important bits of the corresponding block centre pixel that also encoded to DNA using the same rule as in Fig. 3.
5. Apply the DNA-XOR process as shown in Table 2 between the DNA secret sequence and the DNA block centres sequence in which that band will be hidden and converting the resulting code into binary then to pixel bands.
6. Using the LSB technique, the eight bits are hidden after encoding them into the eight bands adjacent to the centre band of that block as shown in Fig. 4. The following is a chart illustrating the embedding procedure as shown in Fig. 5.

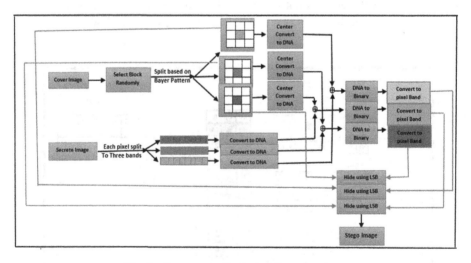

Fig. 5. Show steps for Embedding procedure.

B: Extraction Procedure:

1. Extracting the blocks and their centres in which the hiding the secret image pixels with the same random order and the same initial value used for concealment.
2. Depending on the concept of the Bayer pattern, a block is broken into three blocks as in Fig. 1 to extract each encrypted band of the secret pixels in the block that its band is identical to the block centre band using the LSB technique.
3. Transforming the extracted encrypted secret pixel bands and the centres of their blocks to the DNA code according to Table 1 and using the three least significant bits of the block centre to find the rule number.
4. The DNA-XOR is applied according to Table 2 between the DNA code sequence of the block centre and the DNA code sequence of the extracted pixel bands
5. Converting the DNA sequence obtained from the previous step to the binary using Table 1 and the same rule that depending on the centre of the block to determine the rule (column number in Table 1). Then, convert the binary values to the three bands pixels values that represent the secret image. The following is a chart illustrating the Extraction procedure as shown in Fig. 6.

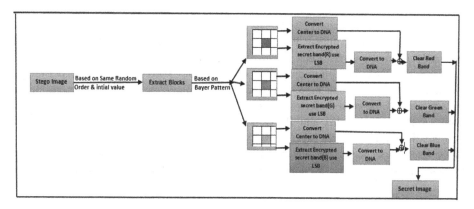

Fig. 6. Show steps of extraction procedure

4 Results

The proposed method is implemented in python 3 using the PyCharm 2021.1.1 version. There are many standard images with different sizes are used to apply the proposed method in several experiments. To evaluate the efficiency of the method, Structural Similarity Index Metric (SSIM) and Peak Signal to Noise Ratio (PSNR) quality metrics are applied between the original and the resulted stego image. The PSNR depends on general image pixels statistics and average calculations and it is a good and efficient quality measure to evaluating the image processing operations [26]. PSNR can be described as in Eq. (1):

$$\mathrm{PSNR(A,B)} = 10\log_{10}\left(\frac{\mathrm{PixelMax}^2}{\frac{1}{X \times Y}\sum_{i=0}^{m-1}\sum_{j=0}^{n-1}(A(i,j) - B(i,j))^2}\right) \tag{1}$$

where PixelMax = 2k-1(k: bits number for each pixel), X and Y: images sizes, A and B: images to be compared. SSIM evaluates the similarity between two images based on luminance, contrast, and structure. In Eq. (2), the formula of SSIM metric:

$$\mathrm{SSIM(A,B)} = \frac{1}{M}\sum_{i=1}^{M}\left(\frac{2\mu_{ai}\mu_{bi} + d_1}{\mu_{ai}^2 + \mu_{bi}^2 + d_1}\right) \times \left(\frac{2\sigma_{aibi} + d_2}{\sigma_{ai}^2 + \sigma_{bi}^2 + d_2}\right) \tag{2}$$

Where, ai, bi are all M windows of A, B images. μ and σ describe the windows mean and standard deviation, respectively. f1 = 0.01, and f2 = 0.03 are default numbers [4]. d1 and d2 are described in Eq. (3) and Eq. (4), respectively.

$$d_1 = (f1 * \mathrm{PixelMax})^2 \tag{3}$$

$$d_2 = (f2 * \mathrm{PixelMax})^2 \tag{4}$$

Where, f1 = 0.01, and f2 = 0.03 are default numbers [4]. Figure 7 illustrated the resulted images for the proposed method steps. These steps are applied on the Lena image on Fig. 7a with size 512 × 512 as a cover and paper image on Fig. 7b with size 170 × 170 as a secret image. Figure 7c shows the center blocks that were selected randomly from the cover Lena image. Figure 7d illustrate the resulted image after applying our DNA encryption method and DNA-XOR between the secret paper image and image in Figs. 7c and Fig. 7e illustrate the resulted stego image using our Bayer pattern method between the cover Lena image and encrypted secret image in Fig. 7d. Table 3 shows the SSIM and PSNR results when applying the proposed method between the original and the stego image with different image sizes for secret and cover image examples. The size of the cover image is larger than the secret image because that each pixel in the secret image is hidden in a 3 × 3 block of the cover image.

Table 3. SSIM and PSNR results for different cover and secret image sizes

Cover image/size	Secret image/size	SSIM	PSNR
Fruits/1000*1000*3	Cat/333*333*3	0.9952	51.6618
Airplane/700*700*3	Pool/233*233*3	0.9957	51.5283
Lena/512*512*3	Watch/170*170*3	0.9963	51.6686
Peppers/256*256*3	Monarch/85*85*3	0.9975	51.6851
Baboon/128*128*3	Tulips/42* 42*3	0.9994	51.7888

From Table 3, we find that the PSNR results are ranged between (51.2–51.7) for all experiments and that is considered acceptable values for statistics measures for resulted images. Also, SSIM results get very good values for all experiments that mean the use of the proposed method is preserve a similar overall structural index between the stego and the original images.

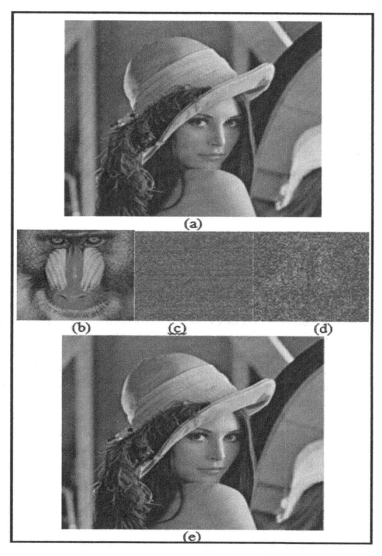

Fig. 7. The proposed method steps resulted image:(a) cover Lena image (b) secret Baboon image (c) random selected center blocks pixels (d) DNA-XOR between DNA encrypted b and c (e) resulted stego image

To prove the efficiency of the proposed method, we compare the proposed method results with other relevant work introduced in Ref. [11]. To get a fair comparison, we apply the same images size for cover and secret images that used in Ref. [11]. We apply our experiment on Lina, Pepper, and Baboon images with size 512×512 and 128×128 for cover and secret image, respectively. Since, we use a large cover image (512×512) with relatively small secret image (128×128), the PSNR metric of our method is increases and it exceeds results presented in Ref. [11] in terms of PSNR metric. Table 4 illustrates the results of experiment when apply the proposed method compared with relevant method in Ref. [11] in terms of PSNR metric.

Table 4. PSNR results of the proposed method applied on Lena, Pepper, and Baboon RGB images with cover size (512×512) and secret size (128×128) compared with method introduced in Ref. [11]

Cover image	PSNR Ref. [11]	Proposed method
Lena	56.87	**57.88**
Pepper	56.27	**57.36**
Baboon	51.96	**53.02**

5   Conclusions

The proposed method aims to hide a color image within a larger color image without major changes in the resulted stego image. The paper adopted a method of randomly selecting the blocks from the cover image to hiding each pixel from the secret image. The secret image and the selected block centers are separated into three color bands and encrypted using DNA rule in special method and apply DNA XOR between the resulted encrypted images. The method hides the secret image in the cover image without using the center pixel of the selected blocks that are used as a public key in the encryption method. The method uses the traditional LSB algorithm after applying the idea of the Bayer pattern to find the locations in three bands that we hide in it. As clear in the results, the proposed method gets a good quality for resulted stego images using SSIM and PSNR quality metrics. The proposed method uses different DNA rules to encrypt each pixel band in secret and selected center blocks images. The rule chosen depends on the least three significant bits in the center blocks image. This step increases the security of the method and adds more complexity and difficulties for the attackers to find the DNA rule for each pixel band. The limitation of the proposed method is the specified size of the secret image with respect to the cover image. Also, since the center of cover image blocks are used as a secret key of the encryption process, it cannot used to hide information in it. As a suggested and future works, Chaotic map and an encryption method as RSA can applied on the secret image before embedding it to increase the security and production.

References

1. Fridrich J, Goljan M.: Practical steganalysis of digital images: state of the art. In: Security and Watermarking of Multimedia Contents IV 2002 Apr 29, International Society for Optics and Photonics 4675, pp. 1–13 (2002)
2. Goth, G.: Steganalysis gets past the hype. IEEE Distrib. Syst. Online **6**(4), 2 (2005)
3. Cox, I.J., Miller, M.L., Bloom, J.A., Fridrich, J., Kalker, T.: Steganography. In: Digital Watermarking and Steganography, pp. 425–467. Elsevier (2008). https://doi.org/10.1016/B978-012372585-1.50015-2
4. Abbood, E.A., Neamah, R.M., Abdulkadhm, S.: Text in image hiding using developed LSB and random method. Int. J. Elect. Comput. Eng. **8**(4), 2091–2097 (2018)
5. Altaay, A.A.J., Sahib, S.B., Zamani, M.: An introduction to image steganography techniques. In: 2012 International Conference on Advanced Computer Science Applications and Technologies (ACSAT), pp. 122–126. IEEE (2012)
6. Reddy, V.L.: Novel chaos based Steganography for Images using matrix encoding and cat mapping techniques. Inf. Secur. Comput. Fraud **3**(1), 8–14 (2015)
7. Valandar, M.Y., Ayubi, P., Barani, M.J.: A new transform domain steganography based on modified logistic chaotic map for color images. J. Inf. Secur. App. **34**, 142–151 (2017)
8. Dash, S., Das, M., Behera, D.: An improved dual steganography model using multi-pass encryption and quotient value differencing. Int. J. Intell. Eng. Syst. **14**(2), 262–270 (2021). https://doi.org/10.22266/ijies2021.0430.23
9. Shreela Dash, M.N., Das, M.D.: Secured image transmission through region-based steganography using chaotic encryption. In: Behera, H.S., Nayak, J., Naik, B., Abraham, A. (eds.) Computational Intelligence in Data Mining. AISC, vol. 711, pp. 535–544. Springer, Singapore (2019). https://doi.org/10.1007/978-981-10-8055-5_48
10. Pradhan, A., Sahu, A.K., Swain, G., Sekhar, K.R.: Performance evaluation parameters of image steganography techniques. In: 2016 International Conference on Research Advances in Integrated Navigation Systems (RAINS), pp. 1–8. IEEE (2016)
11. Ogras, H.: An efficient steganography technique for images using chaotic bitstream. Int. J. Comput. Netw. Inf. Secur. **12**(2), 21 (2019)
12. Bayer, B.E.: Color imaging array, United States Patent 3,971,065, (1976)
13. Zhang, Q., Wang, Q., Wei, X.: A novel image encryption scheme based on DNA coding and multi-chaotic maps. Adv. Sci. Lett. **3**(4), 447–451 (2010)
14. Jiao, S., Goutte, R.: Code for encryption hiding data into genomic DNA of living organisms, In: 2008 9th International Conference on Signal Processing pp. 2166–2169. IEEE (2008)
15. Zhang, J., Fang, D., Ren, H.: Image encryption algorithm based on DNA encoding and chaotic maps. Math. Probl. Eng. **2014**, 1–10 (2014)
16. Bhardwaj, R., Sharma, V.: Image steganography based on complemented message and inverted bit LSB substitution. Proc. Comput. Sci. **93**, 832–838 (2016)
17. Baluja, S.: Hiding images within images. IEEE Trans. Pattern Anal. Mach. Intell. **42**(7), 1685–1697 (2019)
18. Rawat, D., Bhandari, V.: A steganography technique for hiding image in an image using LSB method for 24 bit color image. Int. J. Comput. App. **64**(20), 1–16 (2013)
19. Dogan, S.: A new approach for data hiding based on pixel pairs and chaotic map. Int. J. Comput. Netw. Inf. Secur. **12**(1), 1 (2018)
20. Sun, S.: A novel secure image steganography using improved logistic map and DNA techniques. J. Internet Technol **18**(3), 647–652 (2017)

21. Muhammad, K., Ahmad, J., Farman, H., Zubair, M.: A novel image steganographic approach for hiding text in color images using HSI color model (2015). arXiv preprint arXiv: 1503.00388
22. Hardi, S.M., Masitha, M., Budiman, M.A., Jaya, I.: Hiding and data safety techniques in Bmp image with LSB and RPrime RSA algorithm. J. Phys. Conf. Ser. **1566**(1), 012–084 (2020)
23. Rejani, R., Murugan, D., Krishnan, D.V.: Pixel pattern based steganography on images. ICTACT J. Image Video Proces. **5**(03), 0976–9102 (2015)
24. Qin, X., Tan, S., Li, B., Tang, W., Huang, J.: Image Steganography based on Iteratively Adversarial Samples of A Synchronized-directions Sub-image (2021). arXiv preprint arXiv: 2101.05209
25. Neamah, R.M., Abed, J.A., Abbood, E.A.: Hide text depending on the three channels of pixels in color images using the modified LSB algorithm. Int. J. Elect. Comput. Eng. **10**(1), 809–815 (2020)
26. Rafael, G.C., Richard, W.E., Steven, E.L.: Digital Image Processing. 3rd edn., Eighth Impression (2007)

Panoramic Image Stitching Techniques Based on SURF and Singular Value Decomposition

Nidhal K. EL Abbadi[1] , Safaa Alwan Al Hassani[2(✉)] ,
and Ali Hussein Abdulkhaleq[2]

[1] Computer Science Department, Faculty of Education,
University of Kufa, Najaf, Iraq
nidhal.elabbadi@uokufa.edu.iq
[2] Department of Computer Science, Faculty of Computer Science
and Mathematics, University of Kufa, Najaf, Iraq
safaa.alhassani@student.uokufa.edu.iq

Abstract. The fundamental goal of stitching images is to discover a group of images of a single scene, combine them to form one image for a wide scene. The handheld camera has limited resolution and a narrow field of view. Image stitching faces many challenges such as alignment of image, illumination variation, bad lighting, different scale, and low resolution resulting from stitching. The objective of this paper is to produce a high-resolution panorama in different uncontrolled environments. The suggested method starts by detecting the overlapping areas based on detecting the features extracted by the Speeded-Up Robust Features (SURF) algorithm and Singular Value Decomposition (SVD). Images are automatically aligned based on discovering the geometric relation sips among the images, then images concatenated to create a panoramic image. The suggested algorithm was experimentally tested with different uncontrolled environments such as different resolutions, different image sizes, different numbers of input images, different illumination, and bad lighting. The results showed that the proposed algorithm could stitch correctly with a different overlapping area up to 20% and sometimes more for the images obtained by a wide lens with a different wide-angle. The results were promised and dependable, and the tests prove the ability of the proposed algorithm to solve many image stitching challenges.

Keywords: Panorama · Speeded-up robust features · Singular value decomposition · Image stitching · Image alignment

1 Introduction

Image stitching is a process to combine a sequence of images, mutually having overlapping areas, resulting in a seamless, smooth panoramic image. The handheld camera has limited resolution and small field-of-view, while the image stitching can get high-resolution and high-quality panorama by using handheld equipment. Image stitching has become a hot spot in the field of computer vision, image processing, and computer graphics [1].

© Springer Nature Switzerland AG 2021
A. M. Al-Bakry et al. (Eds.): NTICT 2021, CCIS 1511, pp. 63–86, 2021.
https://doi.org/10.1007/978-3-030-93417-0_5

Aligning and stitching images together is one of the oldest and most widely used techniques in computer vision. Algorithms for image stitching create the high-resolution photos used to create digital maps and satellite images. Additionally, they are included with the majority of current digital cameras and can be used to create stunning ultra-wide-angle panoramas [2]. Since the dawn of photography, combining smaller images to create high-resolution images has been popular [3]. The principle behind image stitching is to combine multiple images into a high-resolution panoramic image based on the overlaps between the images. A specialized form of image known as image stitching has become increasingly common, especially in the making of panoramic images.

To achieve seamless results, there should be nearly exact overlaps between images for stitching and identical exposures. Stitching is not possible if the images do not share a common region. The images of the same scene will have varying intensities, scales, and orientations, and the stitching should work or at the very least produce an output that is visually appealing [5].

The image stitching problem can be divided into two distinct areas of research: image alignment and image stitching. The researchers try to discover the geometric relationships among the images and how the image rotation and the area of images overlapping effects the alignment process. On the other side, working in an uncontrolled environment has a high effect on the panorama resolution and stitching performance. Many other problems affect the stitching, such as blurring or ghosting caused by scene and parallax movement, different image exposures as well as distortions caused by camera lens so that seamless, high-quality panoramas can be achieved [2].

Typically, the image stitching process is divided into three stages: registration, merging, and blending. During image registration, multiple images are compared to determine which translations can be used for image alignment. Following registration, these images are stitched together to create a single image. The purpose of image merging is to obscure the transition between adjacent images visually. In the majority of cases, adjacent image edges exhibit undesirable intensity discrepancies. These intensity variations are present even when the registration of two images appears to be nearly perfect. A blending algorithm is used to eliminate these effects and improve the visual quality of the composite image. The purpose of blending is to determine the final value of a pixel in an area where two images overlap [4].

Several algorithms are introduced to accomplish image stitching. The traditional algorithms perform pixel-by-pixel registration (Direct method), which employs some error criteria to determine the best registration, i.e. the registration with the lowest error value is the best. These methods are slow down and occasionally do not produce the best results. The feature-based registration methods identify distinctive features in each image and then match them efficiently to establish correspondences between pairs of images to determine the approximate motion model, such as SURF&SVD. Feature-based approaches have the advantage of being more resistant to scene movement and potentially faster; when implemented correctly, common feature points are used to establish the relationships between the images that enable automated stitching.

The rest of the paper includes related works in section two, while the types of feature extraction and points are introduced in section three. Section four focuses on the methodology. Section five introduce the results, and finally, section six assigns to the conclusion.

2   Related Works

Several techniques were suggested previously, and many researchers work in the field of image stitching. This section is focused on the reviews of previous methods that have worked in this field of study. Some of them are:

T. Zhang, et al. (2020) The authors proposed an improved SURF algorithm. The feature points are extracted through the Hessian matrix, and then the circular neighborhood of the feature points is used for feature description. Each wave point is used to discover a descriptor for every feature point; Random Sample Consensus (RANSAC) can be used to get rid of all unwanted or unmatched points. In comparison to the conventional SURF algorithm, this algorithm benefits from good speed, full use, and higher accuracy of gray information and detailed information [6]. The accuracy of the stitching was 83.75%, which is considered good, but it needs more time.

Qi, et al. (2019) The authors suggested an improved SURF feature extraction method. The retrieval and recording of images is the core of the image stitching, which contributes directly to the quality of the stitching. Tackling the issue of unequal distribution of images and image stitching. In this proposed, the BRIEF operator in the ORB algorithm performs the rotation shift-invariance. The European pull distance was later used to measure the similarity, and the KNN algorithm is used to calculate the rough features matching. Finally, the larger distance threshold is used to delete the corresponding pair, and the RANSAC algorithm is then used for cleaning. Experiments show that the algorithm proposed is good in real-time but it did not solve the problem of computing in extracting the descriptor of the features, and the mismatch of the similarity of the descriptors, so it is not possible to stitch the multi-images and high-resolution [8].

Shreevastava, et al. (2019) The authors proposed a mismatching point elimination algorithm based on the angle cosine and RANSAC algorithms to address the ORB algorithm's high number of mismatches and low accuracy. Experiments demonstrate that the improved algorithm significantly improves accuracy and is more efficient than the established algorithms SIFT and SURF, making it applicable to various applications requiring high precision and real-time performance. However, it did not solve the problem of stitching the image at the condition of poor lighting, and when there is variation in the geometrical images due to the small number of matching features [9].

M. Wang, et al. (2017) The authors address that the SIFT and SURF features face a problem of long time-consuming. The authors proposed a new image-stitching algorithm using ORB features (Oriented FAST and Rotated BRIEF) to solve the time problem. FAST is used to extract the ORB feature points with directional information, de-scribed by BRIEF. Feature extracting and matching. Also, the false matching points were removed by using the RANSAC algorithm. Lastly, the image blending speeds up by using the weighted average method. The authors proved that image stitching by using the proposed algorithm have the same effect as that of the SIFT and SURF algorithms in terms of results, this method did not solve the problem of illumination variation for the input images, noise, and the difference in the direction of the images, this is due to the instability of rotation [10].

3   Contribution

1. In this method, SVD is used to extract features, and this is the first method used SVD in panorama generation.
2. Suggest a hybrid method that combined SVD and SURF.
3. Reduce time for panorama generates with any number of input images.
4. Generation panorama is robust for image noise, illumination variation, poor illumination.
5. Reduce the effect of the seam line problem as in the previous works.

4   Image Stitching System

The image stitching system accepts two or more input images and outputs the stitched image. The main general steps of image stitching are shown in Fig. 1.

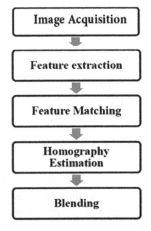

Fig. 1. Block diagram of the general image stitching process

4.1   Image Acquisition

Since the proposed algorithm is based on using the feature-based algorithm, the best features and points must be obtained from the images, and precisely it depends on the specification of the device capturing images. Capturing the image should take care of how to shoot with an angle and view of an entire scene. This helps to get a better result and reduce errors and problems of the resulting panoramic image. Some notes have to be considered when capturing the images. Some of them are:

1. Keep capturing the same scene flat to have the same alignment of the images through a tripod with the lens at its center as in Fig. 2.
2. Images better have the same balance and exposure, especially the white color that represents the lighting, to avoid changing the surfaces of the images and get a good merger between the overlapping areas.
3. Avoid using a lens with an extensive focal length. Therefore, the images will be subjected to high distortion in the lens and may cause alignment problems. As a result, it is recommended to use a focal length of 35 mm or slightly higher.
4. Maintaining the overlap area at 20% or more between every two images as in Fig. 2. The greater the overlap area between the pairs of images, the better the result of stitching will be for the panoramic image.

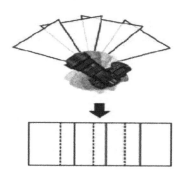

Fig. 2. The process of Image acquiring of four images [19]

4.2 Features Extraction

Features detection is used within computer vision systems to extract features of an image. Detected features are used by motion detection, tracking, panorama stitching, and object recognition system. Points are the simplest features that can be detected. They can be found by corner detection algorithms [14]. The essential characteristics desired in feature extractors are their distinctiveness, accurate and robust keypoint localization, robustness to noise, invariance to orientation, shift, scale, changes in viewpoint, illuminations, occlusions, repeatability, and computationally effectiveness. The most precious property of a feature detector is its repeatability which is the reliability of a detector for finding the same physical feature points under different viewing conditions. The three feature types are Corners, Edges, and uniform intensity regions, as shown in Fig. 3.

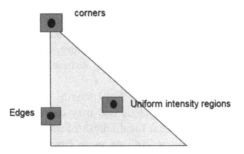

Fig. 3. Type of feature

Finding the correct feature points is critical for performing the correct stitching [12]. It is the first step, and thus it is critical to select the correct detector for this point. Some of these feature detections is:

Singular Value Decomposition

The Singular Value Decomposition is a focus on linear algebra, it gives three matrices as shown in Fig. 4.

$$A = USV^T \tag{1}$$

Where [13].

- U is an m × n matrix with orthonormal columns.
- S is an n × n diagonal matrix with non-negative entries.
- V^T is an n × n orthonormal matrix.

The diagonal values of S are called the Singular Values of A. S is the summation of the diagonal entities λ1, λ2..... which represent the singular vector (called singular values) of a matrix (A) So,

A = λ1 U1 V1^T + λ2 U2 V2^T + ... + λr Ur Vr^T.

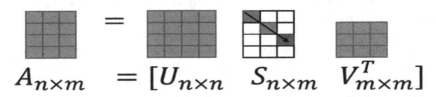

Fig. 4. The resulting matrices from SVD transformation [13]

4.3   Speeded up Robust Features

Speeded up robust features is a scale and rotation invariant interest points detector and descriptor that is used in computer vision works like 3D reconstruction or object recognition. SURF uses a Haar wavelet approximation of the determinant of the Hessian matrix; its descriptor is a collection of Harr wavelet responses around the interesting point. The local area is divided into a 4 × 4 grid. The x and y components of the Harr wavelet response and the absolute value are calculated for each quadrant of the local area. Concatenating these 4 × 4 subregions produce a descriptor vector of length 64 for each interest point. This descriptor is distinctive and robust to noise, detection displacement, and geometric and photometric deformation [14]; it focuses on the scale and in-standard rotation-invariant detectors and descriptors. These seem to offer a good compromise between feature complexity and robustness to commonly occurring photometric deformations [15]. Given a point K = (x, y) in an image I, the Hessian matrix 'H(x, σ) in x at scale σ weighted with a Gaussian is defined as follows[14].

$$H(K, \sigma) = \frac{Lxx(K,\sigma) \quad Lxy(K,\sigma)}{Lxy(K,\sigma) \quad Lyy(K,\sigma)} \tag{2}$$

Where $Lxx(K,\sigma)$ is the convolution of the Gaussian second-order derivative $\frac{\partial 2}{\partial x2}$ g (σ) with the image I in point x, and similarly for $Lxy(K,\sigma)$ and $Lxy(K,\sigma)$.

After extracting the points from the Hessian determinant, the first step for the extraction of the point and orientation along the orientation selected. The size of this window is the 20s [16], where (s) the scale at which the interesting point was detected. Examples of such square regions are illustrated in Fig. 5(A) is preserved important spatial information. For each sub-region, Haar wavelet responses are computed as shown in Fig. 5(B). The dark parts have the weight −1 and the light parts + 1. Orientation assignment: a sliding window orientation of size π/3 detects the dominant orientation of the Gaussian weighted Haar wavelet responses at every sample point within a circular neighborhood around the interesting point, as shown in Fig. 5(C). To build the descriptor, an oriented quadratic grid with 4 × 4 square sub-regions is laid over the interesting point as shown in Fig. 5(D).

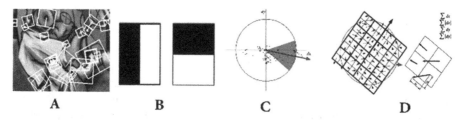

Fig. 5. Type (A) features obtained using Hessian-based detectors, (B) Haar wavelet (C) Orientation assignment (D) To build the descriptor [14]

The primary objective of using integral images in SURF is to reduce the time required to compute key points. They allow the computation of convolution filters of the box type to be achieved in a short amount of time. An integral image denoted I (x, y) at location (x, y) contains the number of the pixel's values above and to the left of (x, y) according to Eq. 3 [14].

$$I(x,y) = \sum_{x',y' \leq x,y} I(x',y').$$ (3)

4.4   Features Matching

Features matching is the process to find corresponding features in two or more different views of the same scene. There are a lot of approaches for matching features. Generally, they can be classified into two categories: region-based matching and feature-based matching. The proposed algorithm will use feature-based matching. After the matching pairs are known, it will return the dependent points to these features and then estimate Geometric Transform [17]. When matching, there will be one feature in the first image and more than one feature in the second image. To solve this problem, we use the unique feature as in Fig. 6.

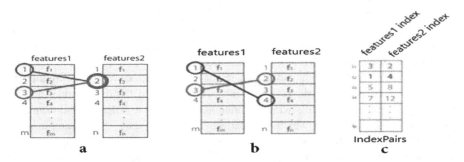

Fig. 6. The matching process: a) matching features1 to features2, b) Choose the unique feature in the case of the similarity of more than one. c) keeps the best match within the identical index pairs

4.5   Homography Estimation

A homography is an invertible transformation from the real projective plane to the projective plane that maps straight lines. The purpose of homography is to specify a point-by-point mapping from the locations of pixels in one image to their corresponding pixels in another image, as illustrated in Fig. 7. This mapping is performed on the assumption that the two images are planar: The two images' points are mapped, assuming that the object being imaged is planar [18]. Two images are related by a homography only if both images are viewing the same plane from a different angle; the Camera may be rotated.

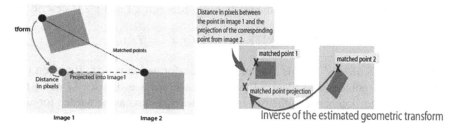

Fig. 7. The process of dropping and dragging points through the space between pixels and the geometry transformation.

Aligning the image requires estimating the geometric transform between each matched pair, which is then used for alignment.

The projection of point X' on a rotating image can be calculated as [19]

$$x' = [K][Rt]X \tag{4}$$

where K is the calibration matrix, R is the rotation matrix, t is the translation vector and X is the world coordinates.

We now have all of the images with a fixed rotational axis, and the position of the P-camera, as shown in Fig. 8, is identical for the acquired images I_1 and I_2.

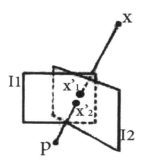

Fig. 8. Aligning two rotational images [19]

As illustrated in Fig. 8, the two images were taken in the same position with no translation; the matching pair can be rewritten as [19]:

$$x'1 = K1R1X \tag{5}$$

$$x'2 = K2R2X \tag{6}$$

By combining Eq. (5) and (6), the relationship between x'1 and x'2 is described as

$$x'2 = K2R2R1TK1 - 1x'1 \; Or \; x'2 = Tx'1 \tag{7}$$

Where $T = K_2R_2 R_1^T K_1^{-1}$ is known as the transformation matrix. Equation (7) is applied to each point of the identical pairs for each point in the I(n-1) and I(n) to estimate the projection between the two images. The equation can be written for all the engineering estimate completely in the form.

$$I(n) = T(n)I(n - 1) \tag{8}$$

The transformation between I (n) and I (n-1) is estimated by T(n).

The hierarchy of 2D linear transformations has several types as shown in Fig. 9, but the most commonly used in panoramas are similarly, affine and projective. The latter is chosen because it maintains collinearity, synchronization, and connection order and is dependent on the maximum number of congruent point pairs and has a minimum of (4), while affine (3), and likewise (2). And the exclusion of non-internal outliers using the RANSAC algorithm.

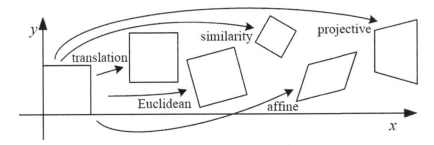

Fig. 9. Different transformations [3]

5   Blending and Composing

The algorithm's final stage is image blending and composition. Before aligning all the images, a 2-D spatial space must be created so that all the images can be aligned in a loop using the transformation matrix. After aligning the images, the image edges must be blended to produce a smooth view. The linear gradient blending technique is used to accomplish this task in this algorithm because it is quicker and produces more accurate results. Gradient blending may be customized to deal with complex alignments that occur when images are rotated or taken in perspective [19]. This is an easy-to-use approach that works well. It produces a linear gradient by shifting the alpha channel from one image's center to another image's center. The term "blend image" is defined as: [19].

$$I_{blend} = I_{Left} + (1-)I_{Right} \tag{9}$$

Where is the gradient weight, which can be any value between 0 and 1, and I_{Left} and I_{Right} are the two images to be blended. After aligning and blending all of the images in the picture package, the panoramic image is eventually composed, as shown in Fig. 10.

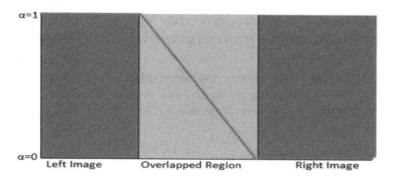

Fig. 10. Gradient blending in the overlapping area, α decreases from 1 to 0 [4].

6 Image Quality Metrics

Image quality measurement (IQM) plays an important role in evaluating the performance of image processing, where the image quality is defined as a property of the image to measure the degradation of the processed image by comparing it to an ideal image.

1. **Normalized Cross-Correlation (NCC):** a tool used to find the relation between two images, normally the source image and the degraded image. Formula to measure NCC is: [20].

$$NCC = \sum_{i=1}^{N} \sum_{j=1}^{m} \frac{(Aij*Bij)}{AIJ^2} \tag{10}$$

Ideal Values = (1), The worst values are greater than (1).

2. **Structure Similarity Index (SSIM):** is defined as a function of luminance comparison, contrast, and structural information of images. The general formula of SSIM is:

$$SSIM(x,y) = [l(x,y)]^{\alpha}.[c(x,y)]^{\beta}.[s(x,y)]^{\gamma} \tag{11}$$

Where α, β, and γ are parameters that define the relative importance of each component, and:

$$l(x, y) = (2\mu_x\mu_y + C1)/(\mu_x^2 + \mu_y^2 + C1) \tag{12}$$

$$c(x, y) = (2\sigma_x\sigma_y + C2)/\left(\sigma_x^2 + \sigma_y^2 + C2\right) \tag{13}$$

$$s(x, y) = (\sigma_{xy} + C3)/(\sigma_x\sigma_y + C3) \tag{14}$$

Where C1, C2, and C3 are constants introduced to avoid instabilities when the average pixel value $(\mu_x^2 + \mu_y^2)$ standard deviation $(\sigma_x^2 + \sigma_y^2)$ or $(\sigma_x + \sigma_y)$ is close to zero. SSIM (x, y) ranges from zero (completely different) to one (identical patches) [20].

1. **Perception-based Image Quality Evaluator (PIQUE):** This tool quantifies the distortion without the need for any training data. It relies on extracting local features for predicting quality. The general formula is [21].

$$PIQUE = \frac{\left(\sum_{k=1}^{N^{SA}} D_{SK}\right) + C_1}{(N_{SA} + C_1)} \tag{15}$$

The quality scale ranges, Excellent (0–20), Good (21–35), Fair (36–50), poor & Bad (51–100).

2. **Natural Image Quality Evaluator (NIQE):** It is used to measure the deviations from statistical regularities observed in natural images, it is a completely blind tool (not need reference image), NIQE formula is [22]

$$NIQE(D_{(v1,v2,\sum 1,\sum 2)})\sqrt{(v1 - v2)^T\left(\frac{\sum 1 + \sum 2}{2}\right)^{-1}(v1 - v2)} \tag{16}$$

NIQE returned a non-negative value. Lower values of score reflect the better perceptual quality of image the input.

3. **Blind/Reference less Image Spatial Quality Evaluator (BRISQUE):** which extracts the point-wise statistics of local normalized luminance signals and measures image naturalness (or lack thereof) based on measured deviations from a natural image model, the general formula is [23]:

$$\hat{I}(i,j) = \frac{I(i,j) - \mu(i,j)}{\sigma(i,j) + C} \tag{17}$$

BRISQUE returned a nonnegative scalar value in the range of [0, 100], A low score value indicates high perceptual quality and a high score value indicates low perceptual quality [23].

7   Methodology

This section explains the proposed stitching method to get a panorama image based on SVD features extraction and the SURF method. The main processes of this proposal are: points detection, matching extracted features of points, and geometric transformation to create a panorama. The image stitching process to produce a clear and high-resolution panoramic image requires a set of steps, whether on a gray or color image as listed in Algorithm (1).

Algorithm 1: SVD-SUFT Stitching
Input: N images, where N number of images.   **Output:** Panorama image.
Begin: 1. Initialize the empty panorama image with width and height (P).   2. Read the first image from the image set.   3. Convert RGB images to grey images.   4. Detection points (poi) of a gray image by the SURF method.   5. Find extract features (fea) of image and points by SURF method to get new points and features.   6. Compute other features by using the SVD algorithm to get U, S, and V matrices of features.   7. Summation SURF features with SVD feathers.   8. Resize the height of the image according to (P).   9. Project the image in the (P).   10. Read the new images.   11. Repeat steps from steps 3 to 7.   12. Map extracted features of the first image with features of the current image using exhaustive feather matching, which relies on the distance between their features. Results are the index of mapped coordinates.   13. Find matched Points of corresponding points in the previous step.   14. Estimate the project transformation for the points from the previous step to the project on the panorama (P).   15. Resize the current image height according to (P).   16. Project the current image in the (P).   17. If there is another image to be stitching, go to step 10.   output= panorama.
End

Step1 creates a zero-empty panoramic matrix, so the images are projected to their specific location. This is done by relying on the dimensions extracted from geometric transformations, and also it depends on the number of images that make the panorama and their sizes. The images are superimposed over others, ending with the last image of the panorama.

The first image will be read from the set of images. The input RGB image is converted into a gray-level image, as well as eliminating any noise and unwanted particles.

Unique and strong points such as the edges of the image or corners will be detected by SURF. Returned a SURF points object, containing information about SURF features detected in a 2-D grayscale. The work is carried out on every image.

After extracting the important and unique points of the image, features are extracted, descriptors are derived from the pixels surrounding the point of interest. They are required to describe and match features defined by a single point location for all image dimensions. When SURF is used to extract descriptors, the orientation property of returned valid_points is set to the orientation of extracted features in radians. This is useful for visualizing the descriptor orientation.

Also, features are extracted by using the SVD transformation. SVD converts any matrix to three matrices (U, S, and V), these three matrices are saved as a vector of features.

Features from SURF and SVD are added to create a new matrix. The new matrix will be useful for Matching and Blending. Image height will be resized according to the empty panorama dimension. The first image is projected on the empty panorama.

The same steps will be repeated for the next image (and each image in the set of images). The matching process at this step will be achieved by the features vector of the current image with the features vector of the previous image. The purpose of this step is to compare the robust features between images and finding a match between the features of the input images. RANSAC algorithm will be used to removes the incorrect points and makes the correct points as the solder points to get a good panoramic image.

The final step in image stitching is the geometric estimation, where the geometric transformation between each matched pair of features is calculated, and estimating the output from the transformation. This step is applied to align all images. The image is inserted in the panorama after resizing its height. Figure 11 shows the block diagram of the suggested algorithm.

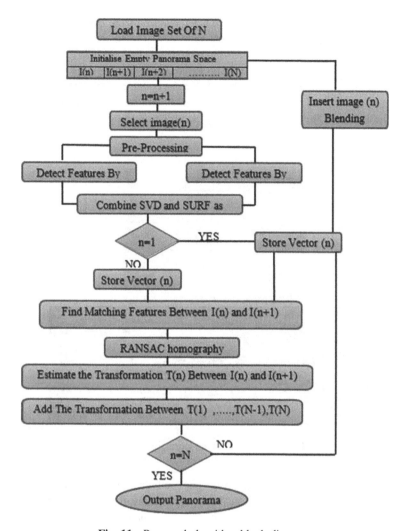

Fig. 11. Proposed algorithm block diagram.

8 Results and Discussion

Many tested are implemented to prove the strength of the proposed algorithm as follow:

8.1 Stitching Images Under Normal Conditions

In this test, images with overlapping regions and without noise or illumination variation are suggested to be used in this test to make a panorama. The number of images used in this test was two images. The result of the stitching process is shown in Fig. 12. Resulted image from this test shows a high-quality panoramic image.

Where images in Fig. 12 (a) are the images acquired from different points, and Fig (b) refers to features extraction and point detection, Fig (c) refers to matching with points, Fig (d) the panoramic images obtained from the suggested method, and Fig (e) The final panorama image after cutting the extreme areas.

The resulting panorama in Fig. 12 (e) and most of the previous algorithms face the problem of appearing a line separating the images that are stitching (bounding in red lines for indication). This problem is due to the variation of lighting of the images to be stitching.

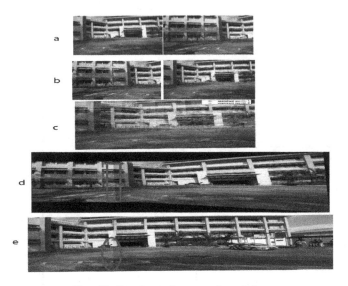

Fig. 12. Results under normal conditions

To solve the problem of appearing lines between stitching images, contrast enhancement will be implemented on the images before the stitching processes. The result of this work show in Fig. 13. The resulting panorama is smooth with high quality.

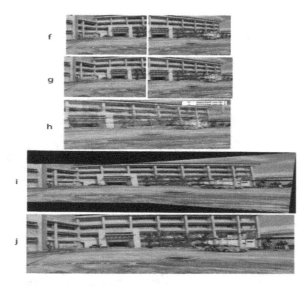

Fig. 13. Results Enhancement under normal conditions

8.2 Stitching Many Images with Normal Conditions

The suggested algorithm was tested by stitching eleven images with a resolution of 1280×1920 pixels without noise or illumination variation. The result of the stitching process is shown in Fig. 14. The resulting image from this test shows a high-quality panoramic image. Panoramic image quality measured by NIQE was 3.34; this value is represented good image quality. The rotation of the camera in this test was $5°$.

Fig. 14. The top row of images is the input images from a synthetic dataset, while the bottom image is the panorama stitching with alignment in the left image, and the panorama stitching with cropping of alignment in the right image.

8.3 Stitching Images with Poor Illumination

The suggested algorithm is also tested by stitching eleven images with a resolution of 5740×3780 pixels, without noise but with poor lighting. The result of the stitching process is shown in Fig. 15. The resulting image from this test shows a high-quality panoramic image. The image quality was measured by several measures, including NIQE, and the result was 3.24; this result indicates good image quality. Also, the image perception measured by BRISQUE, the result was 40.2. This result is also good because the stitching was under poor lighting. The rotation of the camera in this test was 15°.

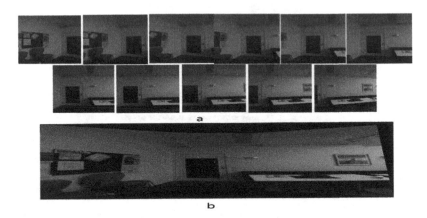

Fig. 15. Show the eleven-input image at the top with the resulted panorama image in the bottom

8.4   Stitching Images with Various Illumination

The strength of the proposed stitching algorithm is tested by stitching three images with a resolution of 2325 × 3150 pixels, without noise, but with various illumination. The result of the stitching process is shown in Fig. 16. The resulting image from this test visually shows a high-quality panoramic image. The image quality was measured by the BRISQUE tool, and the result was 35.25, and this result is good because there are no problems in the image in terms of noise and lighting after the stitching. The rotation of the camera in this test was 20°.

Fig. 16. (a) Images under natural conditions from different angles with stitching output for panorama., (b) Images & panorama after processing using histogram Equalization

8.5   Stitching Images with Noise

The strength of the proposed stitching algorithm is tested by stitching three images with a resolution of 512×512 pixels, with the noise of Gaussian type, and with various illumination. The result of the stitching process is shown in Fig. 17. The resulting image from this test visually shows a high-quality panoramic image. The image quality was measured by using the PIQUE tool, and the result was 24.003, and this result is good and acceptable image quality.

Fig. 17. Show the three-input image with noise at the top with the resulted panorama image in the bottom

8.6   Stitching Images with Geometrical Images

The strength of the proposed stitching algorithm is tested by stitching two images with a resolution of 516×505 pixels, with the geometrical images, the result of the stitching process is shown in Fig. 18. The resulting image from this test visually shows a high-quality panoramic image. Panoramic image quality measured by NIQE was 5.34; this value is represented good image quality. The overlap of the image in this test was 5%

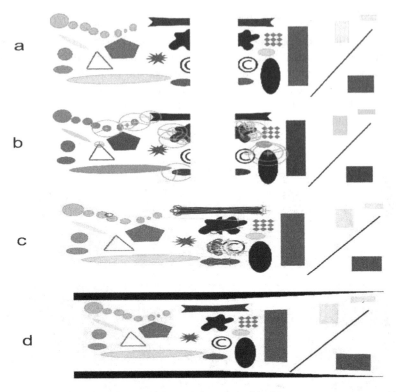

Fig. 18. Results panorama geometrical images

8.7 Comparing the Results

Results of the proposed algorithm were compared visually with other similar works, as shown in Fig. 19. While Table 1 shows the comparison of different methods according to the quality metrics. The results of stitching the panoramic image were uneven.

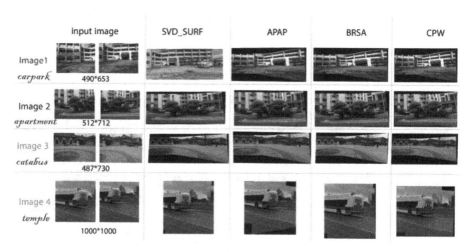

Fig. 19. Results for a collection of techniques used to stitch the images, such as SURF_SVD, APAP and BRSH, CPW for different image dimensions, and different amounts of exposure for several scenes

The input images in Fig. 19 are two images of different dimensions for several scenes.

The algorithms APAP, CPW, BRSH, and the proposed algorithm SURF_SVD used the images in Fig. 19 to create a panorama image and comparing the stitching performance of these algorithms. Each of these algorithms has its stitching mechanism.

Quality assessment tools used to measure the performance of these stitching algorithms results are listed in Table 1. The result shows better performance comparing with the BRSA and CPW and good performance comparing with APAP. Running time or the proposed algorithm is very good comparing with other methods.

Table 1. Measuring image quality after stitching process.

Measures	SVD-SURF	APAP	BRSA	CPW
PIQUE	49.88	43.18	43.38	185.3
NIQE	2.55	1.92	2.64	16.55
BRISQUE	41.26	29.64	47.65	28
NCC	0.86	0.94	0.875	0.82
SIMM	0.43	0.40	0.56	0.39

Figure 20 illustrates the result obtained by applying the proposed method to the apartment dataset [24] with different algorithms, respectively. Each figure is divided into 12 sub-figures denoted by the letters (a) to (l). Each figure's first row (a,c) displays the findings produced using the SIFT technique. Similarly, the second, third, and fourth lines depict the outcomes obtained using the algorithms (SURF-SVD, AKAZE, ORB). Each row contains two columns: the first column contains the initial matching points, while the second column contains the correct matching points after the false match are deleted. The third column shows an image stitched together using a specific algorithm (SIFT, SURF-SVD, AKAZE, ORB).

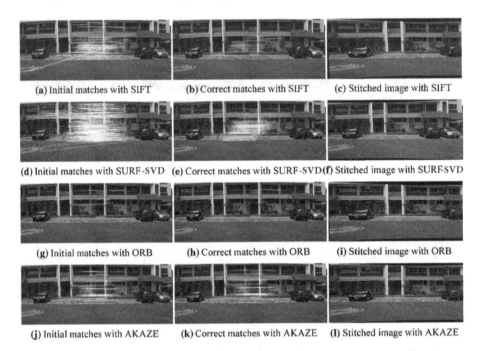

(a) Initial matches with SIFT   **(b)** Correct matches with SIFT   **(c)** Stitched image with SIFT

(d) Initial matches with SURF-SVD   **(e)** Correct matches with SURF-SVD **(f)** Stitched image with SURF-SVD

(g) Initial matches with ORB   **(h)** Correct matches with ORB   **(i)** Stitched image with ORB

(j) Initial matches with AKAZE   **(k)** Correct matches with AKAZE   **(l)** Stitched image with AKAZE

Fig. 20. Results of image stitching and feature matching using the SURF-SVD, ORB, SIFT, and AKAZE algorithms on an apartment dataset

Quality assessment to measure the performance of these stitching algorithms results are listed in Table 2. The result shows better performance comparing with the SIFT and ORB and good performance comparing with AKAZE. Running time or the proposed algorithm is very good comparing with other methods.

Table 2. Comparison of stitched image quality and time required.

Apartment dataset	PSNR	FSIM	VSI	SSIM	TIME
SIFT	15.925	0.6600	0.8724	0.5046	2.1006
ORB	15.403	0.641	0.8640	0.4978	0.1704
SURF-SVD	30.821	0.6676	0.8986	0.5284	1.9271
AKAZE	15.758	0.6733	0.8786	0.5194	1.9759

9   Conclusion

A feature-based image stitching algorithm by SURF_SVD was presented in this paper. Up to our knowledge, this is the first time using the SVD in stitching. The experiment tests prove the robustness and strength of the proposed algorithm against the noise, different resolutions, overlapping size of stitched images, camera focal length, and camera movement. It has a good ability to stitching images in an uncontrol environment. It solves many stitching problems, such as illumination variation and image scaling. It removes the effect of different image illumination and contrast from the resulted panorama. The current algorithm has the ability to stitching many images (more than fifteen images) efficiently. The proposed algorithm is fast and gives a visually very pleasant panorama image. The algorithm compared with other algorithms, and the results were good and promised.

References

1. Kim, H.-K., Lee, K.-W., Jung, J.-Y., Jung, S.-W., Ko, S.-J.: A content-aware image stitching algorithm for mobile multimedia devices. IEEE Trans. Consum. Electron. **57**(4), 1875–1882 (2011). https://doi.org/10.1109/TCE.2011.6131166
2. Kumar, A., Bandaru, R.S., Rao, B.M., Kulkarni, S., Ghatpande, N.: Automatic image alignment and stitching of medical images with seam blending. World. Acad. Sci. Eng. Technol. **65**, 110–115 (2010)
3. Szeliski, R.: Image alignment and stitching: a tutorial. Found. Trends Comput. Graph. Vision **2**(1), 1 (2006). https://doi.org/10.1561/0600000009
4. Abdelfatah, R.A.: Feature-Based Image Stitching. Thesis, pp. 1–71 (2014)
5. Ward, G.: Hiding seams in high dynamic range panoramas. In: Proceedings of the 3rd Symposium on Applied Perception in Graphics and Visualization, p.150 (2006). https://doi.org/10.1145/1140491.1140527
6. Pothula, A., et al.: Application of migration image registration algorithm based on improved SURF in remote sensing image mosaic. IEEE Access **2018**, 163637–163645 (2020). https://doi.org/10.1109/ACCESS.2020.3020808
7. Singh, D.: Multiple images stitching for panoramic image based on RANSAC and homography matrix. Mukt. Shabd. J. **9**(7), 351–359 (2020)
8. Qi, J., et al.: Image stitching based on improved SURF algorithm. In: Yu, H., Liu, J., Liu, L., Ju, Z., Liu, Y., Zhou, D. (eds.) ICIRA 2019. LNCS (LNAI), vol. 11744, pp. 515–527. Springer, Cham (2019). https://doi.org/10.1007/978-3-030-27541-9_42
9. Shreevastava, I., Yerram, K.D., Verma, U.: Image stitching using computer vision algorithms. Digital Transf. Through Indus. **4**, 17–19 (2019)
10. Wang, M., Niu, S., Yang, X.: A novel panoramic image stitching algorithm based on ORB. In: Proceedings of the 2017 IEEE International Conference on Applied System Innovation: Applied System Innovation for Modern Technology, ICASI 2017, pp. 818–821 (2017). https://doi.org/10.1109/ICASI.2017.7988559
11. Adwan, S., Alsaleh, I., Majed, R.: A new approach for image stitching technique using Dynamic Time Warping (DTW) algorithm towards scoliosis X-ray diagnosis. Measurement **84**, 32–46 (2016). https://doi.org/10.1016/j.measurement.2016.01.039
12. Tuytelaars, T., Mikolajczyk, K.: Local invariant feature detectors: a survey. Found. Trends Comput. Graph. Vision **3**(3), 177–280 (2007). https://doi.org/10.1561/0600000017

13. Jia, Y.: Singular Value Decomposition. pp. 1–12 (2020)
14. Bay, H., Ess, A., Tuytelaars, T., Van Gool, L.: Speeded-Up robust features (SURF). Comput. Vis. Image Underst. **110**(3), 346–359 (2008). https://doi.org/10.1016/j.cviu.2007. 09.014
15. MICHAEL KRUIS. Human Pose Recognition Using. Thesis, 1–60 (2010)
16. Viola, P., Jones, M.: MERL-A MITSUBISHI ELECTRIC RESEARCH LABORATORY Rapid Object Detection Using a Boosted Cascade of Simple Features Rapid Object Detection using a Boosted Cascade of Simple Features (2004). http://www.merl.com
17. Muja, M., Lowe, D.G.: Fast matching of binary features. In: Proceedings of the 2012 9th Conference on Computer and Robot Vision, CRV 2012, pp. 404–410 (2012). https://doi.org/ 10.1109/CRV.2012.60
18. Nguyen, T., Chen, S.W., Shivakumar, S.S., Taylor, C.J., Kumar, V.: Unsupervised deep homography: a fast and robust homography estimation model. IEEE Robot. Autom. Lett. **3** (3), 2346–2353 (2018). https://doi.org/10.1109/LRA.2018.2809549
19. Alomran, M., Chai, D.: Feature-based panoramic image stitching. In: 2016 14th International Conference on Control, Automation, Robotics and Vision, ICARCV 2016, vol. 2016, November, pp. 13–15 (2017). https://doi.org/10.1109/ICARCV.2016.7838721
20. Rajkumar, S., Malathi, G.: A comparative analysis on image quality assessment for real time satellite images. Indian J. Sci. Technol. **9**(34), 1–11 (2016). https://doi.org/10.17485/ijst/ 2016/v9i34/96766
21. Venkatanath, N., et al.: Blind image quality evaluation using perception-based features. In: 2015 21st National Conference on Communications, NCC 2015 (2015). https://doi.org/10. 1109/NCC.2015.7084843
22. Mittal, A., Soundararajan, R., Bovik, A.C.: Making a 'completely blind' image quality analyzer. IEEE Signal Process. Lett. **20**(3), 209–212 (2013). https://doi.org/10.1109/LSP. 2012.2227726
23. Moorthy, A.K., Bovik, A.C.: Blind/Referenceless Image Spatial Quality Evaluator. Publisher: IEEE, pp. 723–727 (2011). https://doi.org/10.1109/ACSSC.2011.6190099
24. Zaragoza, J., Chin, T.J., Tran, Q.H., Brown, M.S., Suter, D.: As-projective-as-possible image stitching with moving DLT. IEEE Trans. Pattern Anal. Mach. Intell. **36**(7), 1285–1298 (2014). https://doi.org/10.1109/TPAMI.2013.247

Lossless EEG Data Compression Using Delta Modulation and Two Types of Enhanced Adaptive Shift Coders

Hend A. Hadi[1], Loay E. George[2], and Enas Kh. Hassan[3(✉)]

[1] Ministry of Education, General Education Director of Baghdad Karkh-3, Baghdad, Iraq
[2] University of Information Technology and Communication, Baghdad, Iraq
[3] College of Science\Computer Science Department, University of Baghdad, Baghdad, Iraq

Abstract. Since Electroencephalogram (EEG) signals are treated as datasets, the volume of such datasets are particularly large, EEG compression focuses on diminishing the amount of data entailed to represent EEG signal for the purposes of both transmission and storage. EEG compression is used for eliminating the redundant data in EEG signal. In this paper, a low complexity efficient compression system based on data modulation and enhanced adaptive shift coding is proposed for fast, lossless, and efficient compression. The proposed system starts with delta modulation, which is performed on the raw data, followed by mapping to positive to get rid of any negative values if present, and then two types of enhanced adaptive shift coders are applied. The code-words are passed simultaneously to both optimizers; and the number of bits equivalent to the same code-word from each optimizer is compared, the optimizer with smaller number of bits is chosen for this code-word to be stored in the binary file as the compression outcome. The system performance is tested using EEG data files of Motor\Movement Dataset; the test samples have different size, and the compression system performance is evaluated using Compression Ratio (CR). The experiment outcomes showed that the compression recital of the system is encouraging and outperformed the standard encoder of WinRAR application when it was applied on the same data samples.

Keywords: EEG · Lossless compression · Delta modulation · Mapping to positive · Histogram · Shift coding · Compression ratio

1 Introduction

Electroencephalogram (EEG) is one of the most beneficial type of signals for clinical analysis (i.e., diagnosing diseases and assess the effectiveness of the patent response to medical treatment via the brain functions) [1, 2]. Nonetheless, this analysis procedure habitually takes a vastly long epoch. Considering that each sample of EEG signals is extremely valuable and cannot be pass over, therefore, for the purpose of storage and/or transfer over network. It is necessary that EEG signal should further analyzed using

© Springer Nature Switzerland AG 2021
A. M. Al-Bakry et al. (Eds.): NTICT 2021, CCIS 1511, pp. 87–98, 2021.
https://doi.org/10.1007/978-3-030-93417-0_6

semi lossless methods; as most of the medical data. To compress EEG signals, there are a number of types of data redundancies; such as the temporal redundancy, that should be successfully removed [3, 4].

Many researchers in this field have made comprehensive experiments in order to obtain highest compression ratio with zero loss for the data that these data are sensitive and very important.

Sriraam proposed a lossless compression system for EEG signals neural network analysts based using the notion of correlation dimension (CD). The deliberated EEG signals, as intermittent time series of chaotic procedures, can be categorized as non-linear dynamic parameter CD; which is a degree of the correlation amongst the EEG samples [5].

Wongsawatt et al., presented a method for lossless compression of multi-channel electroencephalogram signals. To exploit the inter-correlation amongst the EEG channels the Karhunen-Loeve transform is employed. The transform is estimated using lifting structure, which outcomes in a reversible awareness under finite precision processing. An integer time-frequency transform is employed for further minimization of the temporal HEOR redundancy [6].

Karimu and Azadi proposed a lossless hybrid EEG compression technique based on the Huffman coding and the property of DCT frequency spectrum. In this technique, the DCT coefficients are computed below 40 Hz (dominant components) of the EEG subdivisions. Formerly, the quantized DCT coefficients are encoded using a Huffman coder in the source site. In the recipient site, to reconstruct EEG subdivisions, a zero set for the DCT coefficients is added above 40 Hz using the inverse DCT [7].

Hejrati et.al, proposed a method that uses intra-channel and inter-channel correlations to suggest a simple and efficient lossless compression technique. In the first stage, a differential pulse code modulation method DPCM is used as a preprocessing phase for extricating intra-channel correlation. Successively, channels are assembled in various clusters, and the centroid of each cluster is computed and encoded by arithmetic coding. Subsequently, the difference between the data of channels in each cluster and the centroid is computed and compressed by arithmetic coding technique [8].

Srinivasan et.al, presented lossless and near-lossless compression system for multichannel electroencephalogram (EEG) signals based on volumetric and image coding. The compression techniques are designed comprehending the concept of "lossy plus residual coding," comprising of a wavelet-based lossy coding layer tailed by arithmetic coding on the residual. Such methodology guarantees an identifiable highest error between reconstructed and original signals [9].

Dauwels et.al, proposed a near-lossless compression system for multi-channel electroencephalogram (MC-EEG) based on matrix/tensor disintegration models. Multi-channel EEG is characterized in proper multi-way (multi-dimensional) structures to effectively utilize spatial and temporal correlations instantaneously [10].

Daou et al., proposed technique for Electroencephalogram (EEG) compression, which exhibits the intrinsic dependence inherent between the diverse EEG channels. It is based on approaches scrounged from dipole fitting that is commonly employed to acquire a solution to the typical problems in EEG analysis: forward and inverse problems. The forward model based on estimated source dipoles is primarily utilized to grant an estimate of the recorded signals to compress the EEG signals. Then, (based on a smoothness factor,) suitable coding procedures are proposed to compress the residuals of the fitting process [11].

Panessai and Abdulbaqi have introduced an effectual system for compressing and transmission of EEG signals based on Run Length Encoding RLE and Discrete Wavelet Transform DWT. Fifty records of EEG signal for different patients were examined from life database. The analysis of these signals can be done in both time and frequency domain (TD) and (FD) respectively under using DWT where it conserves the essential and principal features of the EEG signals. Subsequent step to employ this suggested algorithm is utilizing the quantization and the thresholding over EEG signals coefficients and then encode the signals using RLE that enhances the compression ratio significantly [12].

Titus & Sudhakar proposed a computationally austere and novel approach Normalized Spatial Pseudo Codec (n-SPC) to compress MCEEG signals. Normalization is first applied to the signals followed by two procedures namely the pseudo coding operating on integer part and fractional part of the normalized data and the spatial coding [13].

Many researches proposed the shift coding as an encoding step for the compression system. Hassan et al., proposed color image compression system based on DCT, differential pulse coding modulation, and an adaptive shift coding as encoding step to reduce the bits required to represents the output file [14]. Hashim & Ali, used shift coding optimizer to attain the ideal two code word sizes required to symbolize the small and large succession element values in their image compression system [15]. George et al. proposed a hybrid shift coding to encode the data of selective image encryption system [16].

Ibrahim et al., proposed new high-performance lossy color image compression system using Block Categorization Based on Spatial Details and DCT Followed by Improved Entropy Encoder [17]. Farhan et al., proposed two shift-coding techniques for image compression system, the first technique ("with leading short word") which uses 4 bits to represent a byte which its value less than 15 while the other one indicates the value of the byte to give it either 4 bits or 7 bits [18].

The following is the outline of the article. Section 2 defines the color proposed compression system. Section 3 highlights the metrics used to measure the efficiency of the proposed system. Section 4 contains the experimental findings. Section 5 shows a comparison of the conducted results with WINRAR standard and EDF file format. While Sect. 6 is the conclusion and future suggestions regarding the research area.

2 The Suggested Compression System

The suggested lossless compression system is based on Delta modulation and enhanced adaptive shift coding for EEG data. A public and free dataset is tested in this study; Motor Movement/Imagery dataset contains EEG recordings of 109 healthy volunteers; labeled and depicted in [19]. Each EEG file consist the recordings belong to (64) EEG channel of one volunteer, when the volunteer do some task. The recording duration is ranging from 1 min to 2 min; the EGG recording was sampled at (160 Hz) [19]. In the proposed EEG compression system, the original EEG file is in (EDF) file format; the EEG data is loaded in 2D array. Then it is passed through the proposed compression system, which consist of delta modulation, mapping to positive, and enhanced adaptive shift coding stages, as follows:

Step 1 (applying Delta Modulation DM): DM is the most simple kind of differential pulse-code modulation (DPCM), in which the difference between consecutive samples is encoded to lower the signal values using following equation [16]:

$$DM_{signal}(i,j) = S(i,j) - S(i,j-1) \text{ iff } j > 0 \tag{1}$$

Where DM is the delta modulation matrix of the input signals, S is the signals matrix, j is the signal sample number, and i is the channel number. Where

$$DM_{signal}(i,0) = S(i,0) \tag{2}$$

Equation 2 is applied to preserve the first sample of the input signal to retrieve the original signal in the decompression process [20].

Step 2: The enhanced adaptive shift encoder is applied to reduce the number of bits required to represent the EEG data after the DM step [21]. The encoder is applied to obtain high compression gain throw the following steps:

Applying mapping to positive process on DM array is vital to prevent coding complexity; all DM values are mapped to positive using the following:

$$X_{ij} = \begin{cases} 2X_{ij} & \text{if } X_{ij} > 0 \\ -2X_{ij} - 1 & \text{if } X_{ij} < 0 \end{cases} \tag{3}$$

Where X_{ij} is the element of ith channel and j^{th} sample is registered in the DM array. According to the above equation, the positive values mapped to be even, while the negative values will be odd to distinguish them in the decompression process.

Calculate the histogram of mapped DM array values and find the maximum value to be used in encoder optimizer to determine short code-word, long code-words and tail for reminder [22].

After calculating the histogram of mapped elements values and determine the maximum value in the DM array the adaptive encode optimizer is applied, which is vital to find out the three code-words lengths (short code-word, long code-words, and tail for reminder) to encode the values of the input sequence elements. The code optimizer is enhanced to consider the statistical attributes of the coming stream and makes the proposed system locally adaptive [23].

The optimal values for the code-words lengths should satisfy the criteria "they lead to the minimum total number of bits (T_{bits}) needed to represent all sequence elements values". The enhanced shift coding uses two types of encoding optimizers then only one will be selected adaptively to encode the input file. The optimizer, which led to the minimum number of total bits, is selected.

The first type of the optimizer uses single shift key {0, 1} as key indicators, for short code-word there is no need for key shift, 0 for long code-word type one, and 1 for the reminder. While the second one uses double shift keys {0, 10, and 11} as key indicators for the three code words. The total number of required bits is determined using the following equations [24].

For the first optimizer type (i.e., using shift key)

$$T_{bits} = T_{bits} + His(I) * \begin{cases} n_s & if\ His(I) < R_1 \\ (n_s + n_{l1} + 1) & if\ His(I) < R_1 + R_2 \\ (n_{l1} + n_{l2} + 1) if\ His(I) > R_1 + R_2 \end{cases} \quad (4)$$

Where the array His () is the histogram array, ns, $nl1$ and $nl2$ is the length of short code-word, long code-words, and tail for reminder respectively. $R1$ and $R2$ are determined using the following equations [22]:

$$R_1 = 2^{n_s} - 1 \quad (5)$$

$$R_2 = 2^{n_{l1}} - 1 \quad (6)$$

While the second type of optimizer uses following equation:

$$T_{bits} = T_{bits} + His(I) \times \begin{cases} n_s + 1 & if\ His(I) < R_1 \\ (n_{l1} + 2) & if\ His(I) < R_1 + R_2 \\ (n_{l2} + 2) & if\ His(I) > R_1 + R_2 \end{cases} \quad (7)$$

In Algorithm1, The process of encode optimizer is implemented, where Max is the maximum value in DM array, T_{bits} is the optimal total bits, ns is the short code-word, $nl1$ first long code-word, $nl2$ is the second long code-word for reminder, and OptTyp is the optimizer type (bit key or double bit key). After defining the ideal values of ns, $nl1$ and $nl2$, then the shift coding process is applied to save the output into a binary file as shown in Fig. 1.

```
Algorithm 1:Enhanced Adaptive Shift Coder

Input: DM () array, Max
Output: ns, nl1, and nl2

1. Bt = Log(Max) / Log(2)
2. If 2 ^ Bt < Max Then Bt = Bt + 1
3. For Bt1 = 2 To Bt do
4. R1 = 2 ^ Bt1 - 1: Remm = Max - R1
5. If Remm > 0 Then Btt = Log(Remm) / Log(2)
6. If 2 ^ Btt < Remm Then Btt = Btt + 1
7. For Bt2 = 2 To Btt do
8. R2 = 2 ^ Bt2 - 1: R = Remm - R2
9. If R > 0 Then
10. Bt3 = Log(R) / Log(2)
11. If (2 ^ Bt3<Remm) Then Bt3 = Bt3 + 1
12. Sm = 0
13. For I = 0 To Max do
14. If (I<R1) Then Sm = Sm + His(I) * Bt1
15. ElseIf (I<R1 + R2) Then Sm=Sm+His(I)*(Bt1+Bt2+1)
16. ElseIf (I >= (R1 + R2) Then Sm=Sm+His(I)*(Bt1+Bt3+1)
17. EndFor
18. If(Tbits> Sm)Then Tbits = Sm: OptTyp = 0: ns = Bt1:
  nl1 = Bt2: nl2 = Bt3: EndIf

'''''Shift by double bit key
19. Smm = 0
20. For I = 0 To Max
21. If (I< R1) Then Smm = Smm+His(I)*(Bt1+1)
22. ElseIf(I<R1 + R2)Then Smm = Smm + His(I) * (Bt2 + 2)
23. ElseIf I>=(R1 + R2)Then Smm = Smm + His(I)*(Bt3 + 2)
24. EndFor
25. If Tbits > Smm Then Tbits=Sm: OptTyp=1: ns = Bt1: nl1
= Bt2: nl2 = Bt3: EndIf
26. EndIf
27. EndFor
28. EndFor
```

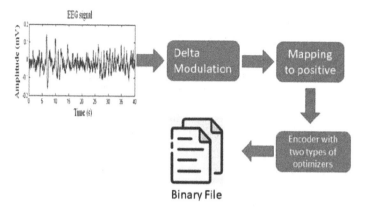

Fig. 1. The proposed compression system

The inverse of the previously mentioned steps is used to decompress and reconstruct the EEG file as shown in Fig. 2.

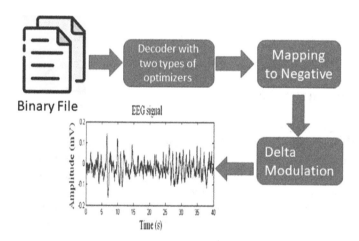

Fig. 2. The decompression steps

3 Evaluation Metrics

The logical and common manner of quantifying a compression algorithm performance is to check the ratio of the number of bits needed to signify the original data to the number of bits needed to signify the compressed data. This ratio is known as the compression ratio [25]. The following metrics have been used in this paper:

3.1   Compression Gain (CG)

The compression gain is calculated as:

$$CG = 10 \times log_{10} \frac{Reference\ Size}{Compressed\ Size} \tag{8}$$

Where the reference size is the size of the input sequence. Where the unit of the compression gain is called percent log ratio and is denoted by percentage [14].

3.2   Compression Ratio (CR)

It is critical to identify how much details can be castoff from the input data so as to conserve significant and vital info of the original data in the procedure of EEG data compression. The Compression Ratio (CR) can be definite as: CR is the amount of diminution of input data size. It can be adapted to get different qualities for the compressed data. The higher the compression ratio the more detailed coefficients to be discarded that might cause lower quality of the reconstructed signal [14], the CR is calculated as:

$$CR = \frac{UnCompressed\ Size}{Compressed\ Size} \tag{9}$$

4   Experimental Results

Many sets of tests are achieved to access the performance of the proposed compression system. The first optimizer type shows compression rates higher than the WinRAR standard application, while the second optimizer type shows acceptable results using different file sizes. Table 1 shows the compression ratio (CR) and compression gain (CG) of the tested files using optimizer type 1 and 2 on the same set of data. To evaluate the performance of each compression system the processing time is another important parameter. Table 2 shows the encoding and decoding time in seconds. The specifications of computer lab top that used to conduct the results are Intel® Core ™ i5-2450M CPU with (4GB) RAM, the operating system is windows-10 (64bit), and the programming language used to develop the compression system is Microsoft Visual C#.

Table 1. The CR and CG of the tested files

File name	Opt. type	CR	CG	File name	Opt. type	CR	CG
S001R01	1	4.91	80%	S035R06	2	3.56	72%
S001R14	1	5.04	80%	S035R11	2	3.58	72%
S002R03	1	4.75	79%	S035R12	2	3.60	72%
S006R12	1	4.78	79%	S050R08	1	5.08	80%
S008R10	1	4.78	79%	S070R05	1	4.77	79%
S010R11	1	4.80	79%	S109R10	1	5.13	81%
S035R01	2	3.56	72%	S035R03	2	3.55	72%

Table 2. The processing time of coding and decoding stages

File	Coding time	Decoding time	File	Coding time	Decoding time
S1R1	0.20	0.08	S35R3	0.50	1.57
S1R14	0.50	0.19	S3R6	0.40	1.49
S2R3	0.60	0.17	S35R11	0.50	1.51
S6R12	0.40	0.18	S35R12	0.50	1.76
S8R10	0.40	0.16	S50R8	0.40	0.16
S10R11	0.40	0.17	S70R5	0.40	0.17
S35R1	0.20	1.62	S109R10	0.40	0.16

5 Comparison with WINRAR and EDF

The attained outcomes of the suggested compression system are competed to the universal standard WINRAR, and the original EDF file format for the signals stored as datasets; regarding compression ratio, compression gain, the results are shown in Tables 3 and 4 respectively. Figure 3 shows the conducted comparison between the proposed system and the standard WINRAR and EDF format.

Table 3. Compressed of file size compared to WINRAR and EDF files

File name	Original data size	EDF file size	Compress file size	WinRAR file size
S001R01	2498560	1275936	508560	687915
S001R14	5120000	2596896	1016208	1400419
S002R03	5038080	2555616	1061507	1288991
S006R12	5038080	2555616	1054293	1275825
S008R10	5038080	2555616	1054293	1293150
S010R11	5038080	2555616	1049758	1610850
S035R01	2498560	1275936	702722	734407
S035R03	5120000	2596896	1442697	1534714
S035R06	5120000	2596896	1441473	1546150
S035R11	5120000	2596896	1431766	1522960
S035R12	5120000	2596896	1425215	1520423
S050R08	5038080	2555616	990792	1240525
S070R05	5038080	2555616	1055281	1401887
S109R10	5038080	2555616	981340	2413408

Table 4. Comparison with WINRAR and EDF files according to compression ratio (CR)

File name	Proposed system CR	WinRAR CR	EDF file CR
S001R01	4.91301	3.63208	1.9582173
S001R14	5.03834	3.65605	1.9715845
S002R03	4.74616	3.90855	1.971376
S006R12	4.77863	3.94888	1.971376
S008R10	4.77863	3.89597	1.971376
S010R11	4.79928	3.12759	1.971376
S035R01	3.55555	3.40215	1.9582173
S035R03	3.54891	3.33613	1.9715845
S035R06	3.55192	3.31145	1.9715845
S035R11	3.57600	3.36187	1.9715845
S035R12	3.59244	3.36748	1.9715845
S050R08	5.08490	4.06125	1.971376
S070R05	4.77416	3.59378	1.971376
S109R10	5.13388	2.08754	1.971376

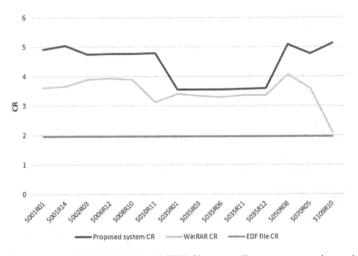

Fig. 3. Comparison with WINRAR and EDF files according to compression ratio (CR)

6   Conclusions and Future Work

In this article, an enhanced EEG compression scheme presented using enhanced adaptive shift coding with two optimizers. For the EEG signal compression, one of the most significant issues is time; the suggested scheme is with minor complexity and short execution time for compression and decompression phases. A good compression ratio achieved when the system applied on some EEG files and compared with WINRAR and EDF files format.

- The use of Delta modulation gives much less values than the original ones to reduce the required number of bits to represent the input.
- Adaptive shift coding modulated using two optimizers to obtain the optimal number of bits to represent the values. When the EEG data entered this stage, the two optimizers calculate the best and least amount of bits needed to characterize the data, the optimizer with least total bits selected to represent the input data in output binary file.

References

1. Hadi, H.A., George, L.E.: Brainwaves for user verification using two separate sets of features based on DCT and wavelet. Int. J. Adv. Comput. Sci. Appl. **9**(1), 240–246 (2018)
2. Dao, P.T., Li, X.J., Do, H.N.: Lossy compression techniques for EEG signals. In: 2015 International Conference on Advanced Technologies for Communications (ATC) (2015)
3. Memon, N., Kong, X., Cinkler, J.: Context-based lossless and near-lossless compression of EEG signals. IEEE Trans. Inf. Technol. Biomed. **3**(3), 231–238 (1999)
4. Sayood, K.: Introduction to Data Compression. Elsevier, USA (2012)
5. Sriraam, N.: Correlation dimension based lossless compression of EEG signals. Biomed. Signal Process. Control **7**(4), 379–388 (2012)
6. Wongsawat, Y., Oraintara, S., Tanaka, T., Rao, K.R.: Lossless Multi-channel EEG Compression. In: IEEE International Symposium on Circuits and Systems, pp. 1611–1614 (2006)
7. Karimu, R.Y., Azadi, S.: Lossless EEG compression using the DCT and the Huffman Coding. J. Sci. Ind. Res. **75**(75), 615–620 (2016)
8. Hejrati, B., Fathi, A., Mohammadi, F.A.: Efficient lossless multi-channel EEG compression based on channel clustering. Biomed. Signal Process. Control **31**, 295–300 (2017)
9. Srinivasan, K., Dauwels, J., Reddy, M.R.: Multichannel EEG compression: wavelet-based image and volumetric coding approach. IEEE J. Biomed. Health Inform. **17**(1), 113–120 (2013)
10. Dauwels, J., Srinivasan, K., Reddy, M.R., Cichocki, A.: Near-lossless multi-channel EEG compression based on matrix and tensor decompositions. IEEE J. Biomed. Health Inform. **17**(3), 708–714 (2013)
11. Daou, H., Labeau, F.: EEG compression of scalp recordings based on dipole fitting. IEEE J. Biomed. Health Inform. **19**(3), 995–1008 (2015)
12. Panessai, Y., Abdulbaqi, S.: An efficient method of EEG signal compression and transmission based telemedicine. J. Theoret. Appl. Inf. Technol. **97**(4), 1060–1070 (2019)
13. Titus, G., Sudhakar, M.S.: A simple but efficient EEG data compression algorithm for neuromorphic applications. IETE J. Res. **6**(33), 303–314 (2018)
14. Hassan, E.K., George, L.E., Mohammed, F.G.: Color image compression based on DCT, differential pulse coding modulation, and adaptive shift coding. J. Theor. Appl. Inf. Technol. **96**(11), 3160–3171 (2018)
15. Hashim, T., Ali, S.A.: Color image compression using DPCM with DCT DWT and Quadtree. Eng. J. **34**(4), 585–597 (2016)
16. George, L.E., Hassan, E.K., Mohammed, S.G., Mohammed, F.G.: Selective image encryption based on DCT, hybrid shift coding and randomly generated secret key. Iraqi J. Sci. **61**(4), 920–935 (2020)

17. Ibrahim, A., George, L.E.: Color image compression system by using block categorization based on spatial details and DCT followed by improved entropy encoder. Iraqi J. Sci. **61**(11), 3127–3140 (2020)
18. Farhan, S., Awad, F.H., Abdulbasit, K., Adeeb, M.: Proposed two shift-coding based compression techniques. In: International Conference on Current Research in Computer Science and Information Technology (ICCIT), Slemani – Iraq (2017)
19. Schalk, G., McFarland, D.J., Hinterberger, T., Birbaumer, N., Wolpaw, J.R.: EEG motor movement/imagery dataset. IEEE Trans. Biomed. Eng. **51**(6), 1034–1043 (2004)
20. Ahmed, H., George, L.E.: Color image compression based on wavelet, differential pulse code modulation and Quadtree coding. Res. J. Appl. Sci. Eng. Technol. **14**(2), 73–79 (2017)
21. Sultan, A., George, L.E.: Image compression based on wavelet, polynomial and Quadtree. J. Appl. Comput. Sci. Math. **5**, 15–20 (2011)
22. Ahmed, H., George, L.E.: Compression image sharing using DCT- wavelet transform and coding by Blackely method. Iraqi J. Comput. Inform. (IJCI) **43**(1), 28–39 (2017)
23. Drweesh, Z.T., George, L.E.: Audio compression based on discrete cosine transform, run length and high order shift encoding. Int. J. Eng. Innov. Technol. (IJEIT) **4**(1), 2277–3754 (2014)
24. Sagheer, M., Farhan, A.S., George, L.E.: Fast intra-frame compression for video conferencing using adaptive shift coding. Int. J. Comput. Appl. **81**(18), 29–33 (2013)
25. Hung, N.Q.V., Jeung, H., Aberer, K.: An evaluation of model-based approaches to sensor data compression. IEEE Trans. Knowl. Data Eng. **25**(11), 2434–2447 (2012)

Automatic Classification of Heart Sounds Utilizing Hybrid Model of Convolutional Neural Networks

Methaq A. Shyaa[1(✉)], Ayat S. Hasan[2], Hassan M. Ibrahim[3], and Weam Saadi Hamza[3]

[1] Department of Information and Communication Systems, Iraqi Ministry of Interior, Baghdad, Iraq
[2] Department of Computer Science, College of Education for Pure Science, University of Diyala, Diyala, Iraq
ayat.saleh.hassan@uodiyala.edu.iq
[3] University of Information Technology and Communication, Baghdad, Iraq
hassan.m@uoitc.edu.iq

Abstract. This paper introduces the employment of a deep learning model to develop an automated classification of heart sound into normal and abnormal. The Cardiology Challenge database 2016 (PhysioNet) has been employed in this work. This dataset is established with the assumption that the phonocardiograms (PCG) are extracted concurrently with the electrocardiograms (ECG). For each heartbeat, the PCG record on a time-domain basis is segmented and is converted to images. Then, these images feed the proposed convolutional neural network (CNN) model to perform the classification. Most previous methods used traditional machine learning techniques to classify the heartbeats. These methods depend on feature extraction and selection methods that are sensitive to different colors, shapes, and sizes. CNN has the ability to achieve feature extraction and classification automatically, unlike traditional machine learning methods which require manual feature extraction, and the performance relies on the chosen features. The proposed is designed by integrating multiple ideas including parallel convolutional layers and residual links. To prove that the proposed model is beneficial in terms of feature extraction and classification, the extracted features have been used to train the SVM classifier. The results show that the model can offer acceptable and comparable accuracy of classification. It has outperformed the previous methods by achieving an accuracy of 93.49% and 94.66% with the SVM classifier. High performance could help id saving many people's lives.

Keywords: Heart sound classification · Deep learning · CNN · PCG

1 Introduction

Most of the excellent achievements in the task of image classification are due to the recent progress in Artificial Intelligence. In the real world, the utilization of deep learning techniques like Recursive Neural Network (RNN), Long-Short Term Memory

© Springer Nature Switzerland AG 2021
A. M. Al-Bakry et al. (Eds.): NTICT 2021, CCIS 1511, pp. 99–110, 2021.
https://doi.org/10.1007/978-3-030-93417-0_7

(LSTM), as well as the Convolutional Neural Network (CNN), generates impressive applications in the fields of biomedical and Bioinformatics [1, 2]. Applications such as breast cancer classification [3] and skin cancer classification [4] are reported in the past three years, as an outcome of CNN algorithm developments, which perform well than the traditional techniques. Heart Sounds have been classified as either normal or abnormal by employing image processing and traditional machine learning methods [5, 6]. These methods depend on feature extraction and selection methods that are sensitive to different colors, shapes, and sizes. Thus, they achieved low performance in the task of heartbeats. On the other hand, deep learning models including CNN have solved the issues of image processing and traditional machine learning methods. Deep learning has shown tremendous performance in a variety of applications [2–4, 7, 8]. One of the best advantages of employing deep learning is the automatic feature extraction and classification unlike machine learning methods [9].

Therefore, this paper introduces a deep learning model for early heart disease detection. Certainly, early, and accurate classification of heartbeats will highly save the lives of the patients. Also, automated monitoring and recording of the ECG and PCG at home, rather than a hospital or clinic, performs as a practical indicative tool for diagnosing any abnormal sign of the heart. Consequently, anybody, who observes any abnormality in his/her ECG or PCG through daily life, could visit a doctor for additional tests. The CNN model proposed has a better feature representation since it uses parallel convolutional neural networks along with the residual connection. We have considered the advantage of previous CNN models. The proposed model was inspired by the design of the model in [3] which has shown impressive performance in different tasks such as diabetic foot ulcer classification [10, 11]. Lastly, to speed up the training process of CNN models, GPU and FPGA have been used to reduce the training time [3, 9, 12]. In this paper, we have utilized GPU to implement the heart sounds classification task. With the recent development in computational tools including a chip for neural networks and a mobile GPU, this work can be extended in the future to be a mobile application. The contributions of this paper can be sum up in two major points (i) novel CNN architecture (ii) Applying a support vector machine classifier (iii) achieving an accuracy of 93.49% and 94.66% with the SVM classifier.

2   Related Work

There were very few techniques or tools based on deep learning available for automatic heart sound diagnosis [13]. The PhysioNet Cardiology Challenge 2016 (abbreviated to "PhysioNet") is the first approach for applying deep learning in biomedical fields (up to the authors' knowledge). However, earlier approaches are based on traditional classifiers of supervised machine learning with pre-extracting feature algorithms. The extracted features from the heart periods are input into support vector machines [14] and Artificial Neural Network (ANN) [15], which are complexity-based features, wavelet features, and time-frequency features. Moreover, earlier works utilized Hidden Markov models in PCG signal segmentation [16] and classification [17].

However, evaluating the success of the earlier works is extremely difficult, due to the variations in the length of the recorded signals, testing algorithms, the available number of recordings for training, dataset quality, as well as the collected data environments. Furthermore, some works did not execute suitable train-test data sets and recorded the results on validating or training data, which is very expected for producing positive results because of overfitting [13]. In addition, to overcome the overfitting problem, similar subject recordings are not involved in both training and validating. A collection of noisy and clean PCG records, which showed extra weak signal quality, is involved in enhancing the development of robust and accurate algorithms.

This paper introduces a deep learning technique, as one of the earliest challenges for heart sound classification. There were several works for applying deep learning techniques to other forms of physiological signal classification. Martinez, et al. [18] described their work, which used deep learning in the Psychophysiology domain. They support the utilization of partiality deep learning for identifying the influence of bodily inputs (e.g., blood volume pulse and skin conductor) inside a study of the game-based user. Besides, they go over the manual use of ad-hoc feature selection and extraction in emotional modeling because it reduces the feature design creativity to them. Furthermore, the reason why this works differently from their work is that they execute a primary unsupervised step of pre-training based on stacked convolutional auto-encoders. In contrast, this work eliminates the need for this step and is trained in an end-to-end supervised fashion instead. Similarly, some works process physical signals using deep learning in human activity diagnosis [19]. PhysioNet 2016 dataset has been used in this paper. Initially, five sub-folders, which are A, B, C, D, and E, are included in the training set of the dataset. Each folder comprises 3126 PCGs. Each PCG record has a time interval of 5 to 120 s [13]. There is the largest number of techniques that transform PCGs into images of exploiting spectrogram methods. For instance, Rubin et al. [20], utilized a logistic regression model based on hidden semi-Markov to segment the beginning of every beat. Next, these beginnings are converted to spectrograms exploiting MFCCs (Mel-Frequency Cepstral Coefficients) method. Later, spectrograms are identified as either normal or abnormal exploiting 2-layer CNN. This CNN has a modified loss function, which augments specificity and sensitivity, accompanied by a regularization parameter. Finally, the last signal classification is the mean probability of the total segment probabilities. This model gets an eighth place at the PhysioNet challenge, with a total score of 83.99%. Kucharski et al. [21], exploited a 5-layer CNN with dropout after transforming the segments by an eight-second spectrogram. This method attained 91.6% in specificity and 99.1% insensitivity. The result is equivalent to the most recent available methods. Another technique by Dominguez et al. [22] segmented each input signal and pre-processing it exploiting the neuromorphic auditory sensor for decomposing the acoustic data into frequency bands. Next, the spectrograms are calculated and fed to a modified AlexNet version. This model attained a vital enhancement contrasted to the prizewinning model of PhysioNet. It achieved an accuracy of 94.16%. Furthermore, Potes et al. [23] exploited Adaboost and CNN. The spectrogram features were fed to Adaboost. Also, cardiac cycles, which were decomposed into four frequency bands, were used for training the CNN. The end outcome is obtained by combining the outputs of the Adaboost and CNN with an overall accuracy of 89%. This model gets the first in the formal aspect of PhysioNet.

On the other hand, models give the impression that has lower performance if there are no conversions from PCGs to spectrograms. For example, Ryu et al. [24] employed a Window-sinc Hamming filter for noise-reduction, signal scaling, and segmentation with a constant window. The used CNN consists of four layers and the achieved accuracy was 79.5%. Shortly later, Chen et al. [25] employed PCG for recognizing the segments S1 and S2. In their work, the PCG signals are converted into a series of MFCCs. Next, the K-means method is applied for clustering the features of the MFCC into two clusters aimed at distinguishing capability as well as refining their representation. Lastly, the SI and S2 features are classified using a DBN. For obtaining the best results, the researchers compared their technique with SVM, logistic regression, Gaussian mixture models, and KNN.

Based on the literature, the most neural networks applied for diagnosing PCG tasks are the CNNs. In addition, like ECG, the spectrogram technique is used for converting PCG signals to images in several deep learning techniques [20, 22, 23] which is the same concept used in this paper. However, the proposed model is more developed than the previous models in aspects of feature extraction and classification.

3   Methodology

3.1   Main Components of CNN

Convolution The organization of the CNN components has an important responsibility in developing innovative structures for attaining improved performance. The following describes the role of CNN components in its structure.

A) Convolutional Layer

This layer consists of a group of convolutional kernels (actually, each kernel is a neuron). The receptive field is defined as a small area of the image, where each kernel is associated with it. Initially, the image is divided into several receptive fields (so-called blocks). Next, each block is multiplying by its corresponding weight (element of filter). This operation can be expressed as in Eq. 1:

$$F_l^k = \left(I_{x,y} * K_l^k\right) \tag{1}$$

Where: $I_{x,y}$ = image input, x,y = spatial locality, and $K_l^k = l^{th}$ convolutional kernel of the kth layer.

The process of image division into receptive fields facilitates in determining locally the values of the correlated pixel. In addition, this locally accumulated information is so-called feature patterns. However, when moving the convolutional kernel over the image along with its weights, various groups of features are extracted. The convolution process can extra be classified into various types based on the path of convolution, the padding type, and the filter size and type.

B) Pooling Layer

As mentioned earlier, the outcome of the convolution process, which is called feature patterns, can happen at various image positions. Since a feature is extracted, its inexact location comparative to others is saved. In contrast, its exact position will not be as important. Downsampling or pooling is similar to convolution as motivating the local process. It accumulates similar information around the reception field and outcomes the main response inside this local area. The pooling process can be expressed as in Eq. 2.

$$Z_l = f_p\left(F_{x,y}^l\right) \tag{2}$$

Where: $Z_l = l^{th}$ output feature map, f_p = pooling process type, and $F_{x,y}^l = l^{th}$ input feature map.

The benefit of the pooling process facilitates the extraction of a feature set, which is unchanged to minor distortions and translational shifts. In addition, it may assist in enlarging the generalization by decreasing the overfitting. Moreover, the decrease in feature map size controls the network complexity. Various pooling formulation types like overlapping, L2, average, and max-pooling are employed to extract translational unchanged features.

C) Activation Function

This function assists in learning a complex pattern and works as a decision function. In general, selecting a proper activation function accelerates the learning operation. Equation 3 defines the function of the convolved feature map.

$$T_l^k = f_A\left(F_l^k\right) \tag{3}$$

Where: T_l^k = transformed output for k^{th} layer, f_A = activation layer, and F_l^k = output of a convolution process.

From literature, several activation functions like ReLU, max-out, tanh, sigmoid, as well as, ReLU alternatives like PReLU, ELU, and leaky ReLU are utilized for training nonlinear feature sets. However, ReLU and its alternatives are chosen, since it facilitates solving the problem of vanishing gradient.

D) Batch Normalization

It is employed for addressing the problems associated with the internal covariance shift inside the feature maps. This internal shift can be defined as a variation in the distribution of the values of the hidden units, which imposing the rate of learning to a lower value (i.e., slowing down the convergence) and necessitates well-thought-out parameter initialization. Equation 4 represents the batch normalization of the transformed feature map.

$$N_l^k = \frac{T_l^k}{\sigma^2 + \sum_l T_l^k} \tag{4}$$

Where: N_l^k = normalized feature map, T_l^k = input feature map, and σ = variation in a feature map.

However, batch normalization brings together the value distribution of the feature map via carrying these values to zero mean and unit variance. In addition, it makes the gradient flow smoother and works as a regulating factor, which enhances the network generalization with no dependence on dropout.

E) Dropout

It presents regularization inside the network. It eventually enhances generalization via arbitrarily moving over some connections and units including a particular probability. Multi-connection occasionally becomes co-adapted and may produce overfitting if it learned non-linear relation. In contrast, various thinned network architectures are produced due to the randomly dropping of some units or connections. One of these architectures is selected with small weights. It is considered a representative network, which represents an approximation of the whole of the intended networks.

F) Fully Connected Layer

It is typically utilized for classification purposes as a last layer in the network. In general, it receives and analyzes the output of all preceding layers. Furthermore, it generates a non-linear grouping of the chosen features, which are employed for the data classification. Unlike convolution and pooling, it is a universal task.

3.2   The Proposed Model

The proposed model is designed to have better feature representation and classification of the heart sound. It starts by employing traditional convolution layers to reduce the size of input images. The first traditional convolutional has a filter size of 5 × 5. Then it is followed by batch normalization and ReLU layers to speed up the training process and to avoid gradient vanishing issues. Both batch normalization and ReLU layers come after every single convolutional layer in the model. The second traditional convolutional has a filter size of 7 × 7. After the traditional convolution layers, three blocks of parallel have been employed to extract good features. Each block consists of three branches. Each branch has a convolutional layer followed by batch normalization and ReLU layers. The first convolutional layer in the first branch of all three blocks has a filter size of 3 × 3. The second convolutional layer in the second branch of all three blocks has a filter size of 5 × 5. The third convolutional layer in the third branch of all three blocks has a filter size of 7 × 7. The output of three branches is combined in the concatenation layer then it pushes to the next block. We have also used residual connections in blocks one and three for better feature representation. At the end of the model, an average pooling layer is added to reduce the dimensionality and decrease the effect of overfitting. Two fully connected layers have been added and between them, the dropout layer is to avoid overfitting issues. All heart sounds converted to signal images in the size of 512 × 512. We have divided the dataset into 70% for training and 30% for testing. The model has been trained with 100 epochs using MatLab as software and GPU as hardware. Figure 1 shows an example of a learned filter by the first convolutional layer while Fig. 2 shows the design of the proposed model.

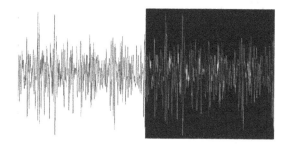

Fig. 1. Example of the learned filter from the first convolutional layer.

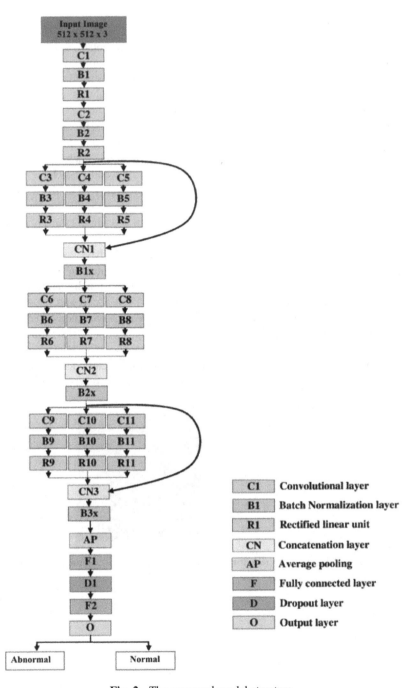

Fig. 2. The proposed model structure.

4 Experimental Results

We have assessed the proposed models in terms of sensitivity, specificity, precision, MAcc as described in [11]. We have worked on PhysioNet 2016 datasets as mentioned earlier. We have compared the proposed model with previous methods that used the same dataset as listed in Table 1. It outperformed the previous methods by achieving a sensitivity of 89.51%, specificity of 97.48%, and MAcc of 93.49%. To improve the results further and to prove the ability of the proposed model in terms of feature extraction, we have used the extracted features by the proposed model to train the SVM classifier. the proposed model with SVM has achieved 91.44%, 97.89%, 94.66% for sensitivity, specificity, MAcc, respectively. The proposed model with the SVM classifier has improved the results due to the good features extracted by the proposed model.

Table 1. Performance comparison between methods using the same dataset.

Methods	Sensitivity (%)	Specificity (%)	MAcc (%)
Potes et al. [23]	94.24	77.81	86.02
Zabihi et al. [26]	86.91	84.90	85.90
Plesinger et al. [27]	76.96	91.25	84.11
Rubin et al. [20]	72.78	95.21	83.99
Langley et al. [28]	94%	65%	80%
Krishnan PT et al. [29]	87%	85%	86%
The proposed model	**89.51**	**97.48**	**93.49**
The proposed + SVM	**91.44**	**97.89**	**94.66**

We have tested some heart sound with the proposed model and effectively classified them correctly.

5 Conclusion

This paper presented a hybrid CNN model that combining multiple ideas including parallel convolutional layers and residual links for the task of automated classification of heart sound into normal and abnormal. The parallel convolutional layers have different filter sizes to obtain better feature representation. PhysioNet 2016 has been used in this work which is a very challenging dataset. Each heartbeat sound was converted to images then these images were utilized to train the proposed model. The proposed model has shown excellent results by achieving an accuracy of 93.49%. Furthermore, we have utilized the features that have been extracted by the proposed model to train the SVM classifier. The proposed model with SVM has achieved an accuracy of 94.66% which outperformed the previous methods. The proposed model has proved that it is effective in terms of feature extraction and classification. The plan is to use it to classify ECG heartbeats sound with the employment of transfer learning plus build it as a mobile application. as shown in Fig. 3.

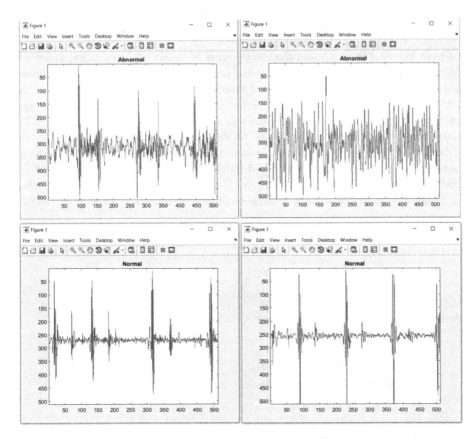

Fig. 3. Heart sound is classified by the proposed model. The first row is abnormal cases, the second row is normal cases.

References

1. Li, Y., Huang, C., Ding, L., Li, Z., Pan, Y., Gao, X.: Deep learning in bioinformatics: Introduction, application, and perspective in the big data era. Methods **166**, 4–21 (2019)
2. Alzubaidi, L., Fadhel, M.A., Al-Shamma, O., Zhang, J., Duan, Y.: Deep learning models for classification of red blood cells in microscopy images to aid in sickle cell anemia diagnosis. Electronics **9**(3), 427 (2020)
3. Alzubaidi, L., Al-Shamma, O., Fadhel, M.A., Farhan, L., Zhang, J., Duan, Y.: Optimizing the performance of breast cancer classification by employing the same domain transfer learning from hybrid deep convolutional neural network model. Electronics **9**(3), 445 (2020)
4. Pacheco, A.G.C., Krohling, R.: An attention-based mechanism to combine images and metadata in deep learning models applied to skin cancer classification. IEEE J. Biomed. Health Inform. **25**, 3554–3563 (2021)
5. Fadhel, M.A., Al-Shamma, O., Oleiwi, S.R., Taher, B.H., Alzubaidi, L.: Real-time PCG diagnosis using FPGA. In: Abraham, A., Cherukuri, A.K., Melin, P., Gandhi, N. (eds.) ISDA 2018 2018. AISC, vol. 940, pp. 518–529. Springer, Cham (2020). https://doi.org/10.1007/978-3-030-16657-1_48

6. Alshamma, O., Awad, F.H., Alzubaidi, L., Fadhel, M.A., Arkah, Z.M., Farhan, L.: Employment of multi-classifier and multi-domain features for PCG recognition. In: 2019 12th International Conference on Developments in eSystems Engineering (DeSE), pp. 321–325. IEEE, October 2019
7. Hasan, R.I., Yusuf, S.M., Alzubaidi, L.: Review of the state of the art of deep learning for plant diseases: a broad analysis and discussion. Plants 9(10), 1302 (2020)
8. Alzubaidi, L., Fadhel, M.A., Al-Shamma, O., Zhang, J., Santamaría, J., Duan, Y.: Robust application of new deep learning tools: an experimental study in medical imaging. Multimed. Tools Appl. 1–29 (2021)
9. Alzubaidi, L., et al.: Review of deep learning: concepts, CNN architectures, challenges, applications, future directions. J. Big Data 8(1), 1–74 (2021)
10. Alzubaidi, L., et al.: Towards a better understanding of transfer learning for medical imaging: a case study. Appl. Sci. 10(13), 4523 (2020)
11. Alzubaidi, L., et al.: Novel transfer learning approach for medical imaging with limited labeled data. Cancers 13(7), 1590 (2021)
12. Al-Shamma, O., Fadhel, M.A., Hameed, R.A., Alzubaidi, L., Zhang, J.: Boosting convolutional neural networks performance based on FPGA accelerator. In: Abraham, A., Cherukuri, A.K., Melin, P., Gandhi, N. (eds.) ISDA 2018 2018. AISC, vol. 940, pp. 509–517. Springer, Cham (2020). https://doi.org/10.1007/978-3-030-16657-1_47
13. Liu, C., et al.: An open access database for the evaluation of heart sound algorithms. Physiol. Meas. 37(12), 2181 (2016)
14. Ari, S., Hembram, K., Saha, G.: Detection of cardiac abnormality from PCG signal using LMS based least square SVM classifier. Expert Syst. Appl. 37(12), 8019–8026 (2010)
15. Uğuz, H.: Adaptive neuro-fuzzy inference system for diagnosis of the heart valve diseases using wavelet transform with entropy. Neural Comput. Appl. 21(7), 1617–1628 (2012)
16. Springer, D.B., Tarassenko, L., Clifford, G.D.: Logistic regression-HSMM-based heart sound segmentation. IEEE Trans. Biomed. Eng. 63(4), 822–832 (2015)
17. SaraçOğLu, R.: Hidden Markov model-based classification of heart valve disease with PCA for dimension reduction. Eng. Appl. Artif. Intell. 25(7), 1523–1528 (2012)
18. Martinez, H.P., Bengio, Y., Yannakakis, G.N.: Learning deep physiological models of affect. IEEE Comput. Intell. Mag. 8(2), 20–33 (2013)
19. Inoue, M., Inoue, S., Nishida, T.: Deep recurrent neural network for mobile human activity recognition with high throughput. Artif. Life Robot. 23(2), 173–185 (2017). https://doi.org/10.1007/s10015-017-0422-x
20. Rubin, J., Abreu, R., Ganguli, A., Nelaturi, S., Matei, I., Sricharan, K.: Recognizing abnormal heart sounds using deep learning. arXiv preprint arXiv:1707.04642 (2017)
21. Kucharski, D., Grochala, D., Kajor, M., Kańtoch, E.: A deep learning approach for valve defect recognition in heart acoustic signal. In: Borzemski, L., Świątek, J., Wilimowska, Z. (eds.) ISAT 2017. AISC, vol. 655, pp. 3–14. Springer, Cham (2018). https://doi.org/10.1007/978-3-319-67220-5_1
22. Dominguez-Morales, J.P., Jimenez-Fernandez, A.F., Dominguez-Morales, M.J., Jimenez-Moreno, G.: Deep neural networks for the recognition and classification of heart murmurs using neuromorphic auditory sensors. IEEE Trans. Biomed. Circuits Syst. 12(1), 24–34 (2017)
23. Potes, C., Parvaneh, S., Rahman, A., Conroy, B.: Ensemble of feature-based and deep learning-based classifiers for detection of abnormal heart sounds. In: 2016 Computing in Cardiology Conference (CinC), pp. 621–624. IEEE, September 2016
24. Ryu, H., Park, J., Shin, H.: Classification of heart sound recordings using convolution neural network. In: 2016 Computing in Cardiology Conference (CinC), pp. 1153–1156. IEEE, September 2016

25. Chen, T.E., et al.: S1 and S2 heart sound recognition using deep neural networks. IEEE Trans. Biomed. Eng. **64**(2), 372–380 (2016)
26. Zabihi, M., Rad, A.B., Kiranyaz, S., Gabbouj, M., Katsaggelos, A.K.: Heart sound anomaly and quality detection using ensemble of neural networks without segmentation. In: 2016 Computing in Cardiology Conference (CinC), pp. 613–616. IEEE, September 2016
27. Plesinger, F., Jurco, J., Jurak, P., Halamek, J.: Discrimination of normal and abnormal heart sounds using probability assessment. In: 2016 Computing in Cardiology Conference (CinC), pp. 801–804. IEEE, September 2016
28. Langley, P., Murray, A.: Heart sound classification from unsegmented phonocardiograms. Physiol. Meas. **38**(8), 1658 (2017)
29. Krishnan, P.T., Balasubramanian, P., Umapathy, S.: Automated heart sound classification system from unsegmented phonocardiogram (PCG) using deep neural network. Phys. Eng. Sci. Med. **43**(2), 505–515 (2020). https://doi.org/10.1007/s13246-020-00851-w

Hybrid Approach for Fall Detection Based on Machine Learning

Aythem Khairi Kareem[1] and Khattab M. Ali Alheeti[2(✉)]

[1] Department of Heet Education, General Directorate of Education in Anbar,
Ministry of Education, Hit, Anbar 31007, Iraq
aytl9cl004@uoanbar.edu.iq
[2] College of Computer Science and Information Technology,
University of Anbar, Ramadi, Iraq
co.khattab.alheeti@uoanbar.edu.iq

Abstract. The number of older people living independently in their own homes is growing worldwide due to the high cost of health care services. Accordingly, it is critical to developing an accurate system with the intelligence to detect human falls during daily activities. In this study, an intelligent system is suggested to effectively detect the fall from daily life activities and to reduce the false alarm. This system employed Machine Learning (ML) techniques because of its classification capabilities. The proposed method is tested and evaluated based on a publicly available University of Rzeszow Fall Detection dataset (URFD). This dataset contains two different signals, namely depth maps from Kinect camera and tri-axial data from ADXL345 accelerometer.

The machine learning techniques used in this system are Support Vector Machine (SVM), Naïve Bayes (NB), and Decision Tree (DT). By applying a support vector machine to a subset of features, the experimental results show that the accuracy is 99.96, with an execution time of 0.272 s. The results are obtained by applying the Naïve Bayes is 98.1 accuracy rate with 0.168 s of execution time. The decision tree technique results are 99.96 rates of accuracy with 0.771 s of execution time.

Keywords: Decision Tree · Naïve Bayes · Support Vector Machine

1 Introduction

Fall, especially in older people, is a major health problem worldwide. Reliable FD systems can decrease the negative consequences of falls. The population of elderly adults over 65 in developed countries has drastically grown in the latest few years. Most elderly adults favor staying in their own houses rather than moving to care facilities because of expensive healthcare costs and privacy concerns. Studies have shown that 30% of the elderly adults over 65 have at least one fall each year and that 47% of elderly adults who have fallen cannot then get back up without support from others [1, 2]. Hence, different approaches of FD have been introduced to detect falls. These approaches utilize various methods to detect falls, such as deep learning systems and Machine Learning (ML), and to use multiple devices to monitor elderly people.

© Springer Nature Switzerland AG 2021
A. M. Al-Bakry et al. (Eds.): NTICT 2021, CCIS 1511, pp. 111–130, 2021.
https://doi.org/10.1007/978-3-030-93417-0_8

Currently, various sensors kinds such as wearable sensors and cameras, can be utilized to monitor a person's body motion or pose to detect elderly people falls without interfering in their routines of life [3]. After using a Fall Detection (FD) system, the researchers explained that people obtained more trust and independence. Accordingly, several technologies have been modified to design, implement and expand techniques for detecting falling in an unsupervised context [4] (Fig. 1).

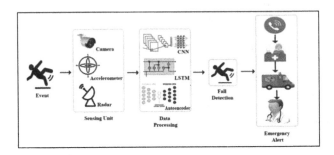

Fig. 1. Fall detection system [3].

The above Figure includes a sensing unit (camera, Radar, and accelerator), data processing (to distinguish the fall event from ADL), and emergency alert.

Three main approaches for FD systems depend on whether data is obtained with vision devices, wearable sensors, or ambient sensors. On another hand, fall detectors can be classified into two broad methods: context-aware systems and wearable devices. Context-aware methods reflect all systems relating to sensors used in the environment, including microphones, infrared, radar, floor, pressure sensors, and vision-based devices. Kinect, Cameras, and motion obtain devices are also estimated context-aware systems. The wearable-based sensors approach utilizes specific devices placed on the body of the elderly person to detect any modification of his movements. Various wearable sensors are utilized for FD, such as smartwatch, gyroscope, and accelerometer [2].

Generally, fall detection schemes either utilize ML or a threshold approach. Machine learning (ML) has appeared concurrently with high performance and big data technologies and computing to design increased possibilities to unravel, quantify, and understand intensive data methods in different environments. ML is known as the scientific domain that provides machines the capability to determine without being surely programmed [5].

In this paper, FD system is proposed; the primary objective of this study has been developed the inexpensive computational system and fall hazard detection. The analysis of this study is performed by using vision and wearable devices. A ML technique has been selected to succeed the disadvantages of threshold approaches generally applied in fall systems. These approaches are SVM, NB, and DT. In addition, it has the ability to decrease the computational complexity of the intelligent classifier.

The work is organized into the following sections: In Sect. 2, Related Works are displayed. In Sect. 3, the overview of machine learning is described. Section 4 explains the fall detection system. Section 5 describes the evaluation metrics. In Sect. 6, the experiment results are presented. In Sect. 7, the comparison of classification techniques is described. Finally, Sect. 8 includes the conclusions and future works.

2 Related Work

This section sponsors a review of previous studies on fall detection and development techniques to enhance the classification algorithms' performance. The literature survey concerning the fall aspects and the survey results are presented below:

Thiago et al. [6] presented two different approaches. Threshold-based algorithms were the first related. This method was achieved by calculating the acceleration by using Madgwick's decomposition, and this data was combined with system velocity. Moreover, they were leading to 91.1% accuracy. The second approach is a ML method, the K-Nearest Neighbor (k-NN) process was the best result where the accuracy was 99%.

José Alves et al. [7] provided one wearable device accelerometer algorithm for three sites of the body: chest, waist, and pocket, without the requirement to have a calibration step. This algorithm can be executed absolutely for data processing, wearable devices, and no external devices are needed. In addition, an analysis of the sample rate of the accelerometer was performed to improve the algorithm's efficiency. Continuous data collection has validated the algorithm. By using the sensor in the user's waist, the suggested algorithm was able to make up for reducing false alarms and increasing the FD score medium algorithm sensitivity point.

Sheikh Nooruddin et al. [8] presented FD and rescue system depended on the Internet of Things (IoT). This system can use different devices such as Arduino, Smartphone, embedded system, and Raspberry Pi to monitor many peoples. These devices can be placed in the users' pockets. The data send continuously from the devices to the server, which comprises the model of ML to analyze the data for discovering the fall. The proposed scheme reached 96.3% sensitivity, 99.6% specificity, and 99.7% accuracy.

J. Jeffin Gracewell et al. [9] introduced an approach to analyzing FD by combining two methods as one system. A two SVM stage classifier distinguishes fall events from ADL. These methods are handling various temporal and spatial features. This system is very efficient, with a decreased error rate of classification. The key frames selection can be depended on the center point of displacement of the target detection that possesses a threshold more significant than the predefined value.

Eduardo Casilari et al. [8] investigated the capability to use deep learning to distinguish falls from ADL based on data taken from sensors. Whether it is a gyroscope or an accelerometer, one of CNN's advantages is that it avoids complex processing, and the deep learning can extract raw data that taken directly from sensors. This feature increases the ability to distinguish falls from ADL detection.

Xugang Xi et al. [10] proposed a study based on signals of plantar pressure and surface electromyography (EMGs). Signals of plantar pressure were collected and extracted features. Features were combined and selected for fall, gait, and posture transition using the fisher class's separability index. This method named as canonical of global analysis of correlation of weighting genetic algorithm. Weighted Kernel Fisher

Linear Discriminant Analysis (WKFDA) was present fall and classification. Results showed that the accuracy of classification of gait and fall reach 98%, and the transition of posture was 100%.

Faten A. Elshwemy et al. [11] proposed a two-stage framework: preprocessing of data deep learning model. The structure is applied to flow dense optical to extract the motion information in the first stage. In the second stage, the deep learning model depending on the Convolutional Long Short-Term Memory Auto Encoder (ConvLSTMAE) network used to extract video temporal and spatial features. An auto encoder reconstruction error is used to recognize falls. The experimental results determine that the framework high performance falls detection compared to models of other deep learning.

R. Jansi et al. [12] proposed a framework to detect falls by using two different tri-axial data. Signals are obtained from depth maps of a Kinect sensor and accelerometer. Firstly, the accelerometer data is monitored continuously and applied to show a fall when the data sum vector quantity from tri-axial meets a particular threshold. The approach suggests a new Entropy, a descriptor of depth difference gradient map, that discriminates fall from ADL. Finally, confirmation of fall by employing classifier by using the extracted descriptors. The results of experimental depict the high performance of this model.

Bohua Wang et al. [13] proposed a method that divided the event of fall into two divisions: fallen-state and falling-state, which depicts the events of fall from static and dynamic aspects. First, the model of object detection You Only Look Once (Yolo), and the model of the detect posture of human Open Pose are utilized for preprocessing to get a human body the position information and key points. Then, the model of a dual-channel sliding window is produced to obtain the human body's static features and dynamic features (upper limb velocity, centroid speed). Next, random forest and Multilayer Perceptron (MLP) are used to classify static and dynamic feature data individually. The results of the classification are fused for FD. Test results present the method reaches an accuracy of 97.33%.

Yuya Ogawa et al. [14] presented a detection of fall system utilizing infrared array sensors. The system provides for the detection of falls that is cheap and suitable for privacy protection in a non-wearable mode. In this system, temperature distribution analyzes utilizing ML to allow more accurate and quicker detection of falls. Then, ML algorithms are used in classifiers to compare and calculate accuracy. Some of the data are a set of data of temperature for two seconds. Some temperature distribution can be gained every 0.1 s by using sensors of the infrared array. The system achieves 97.75% accuracy.

Abdulaziz Alarifi et al. [15] proposed an efficient and optimized FD system that employ a technique based on killer heuristics optimized AlexNet CNN. Wearable sensor methods, consisting of a gyroscope, magnetometer, and accelerometer, are attached to the person's body in six various locations. Through the data acquisition process, 20 falls, and 16 ADL information are obtained in 2520 experiments. Data is obtained from the IoT wearable equipment for feature extraction and analysis of data of sensors. The features are analyzed by multilinear essential element analysis, which decreases the features' dimension. The FD based on the created wearable sensor device is estimated by applying simulation results, and the system identifies a fall with the highest accuracy and lowest complexity.

Gordon Morison et al. [16] suggested an FD system based on Hybrid Multichannel Random Neural Network (HMCRNN) architecture. This system employs raw signals accelerometer data to produce a high accuracy ratio of classification. HMCRNN performs more reliable than the MCRNN, with enhancement of 20%. HMCRNN implementation allows high accuracy rate of 92.23% for FD, comparable to the CNN performance of 91.22% at a significantly lower computational cost.

Our work distinguishes the fall from other activities by applying a machine learning technique. However, this work employed SVM, NB, and DT techniques by utilized a set of features. These features are extracted from Kinect camera and accelerometer sensor. The system aim is to increase the accuracy ratio and at the same time to decrease the executing time and false alarms rate.

3   An Overview on Machine Learning

In recent years, with the speedy evaluation of communication, big data, and popular Internet technologies. The resources, infrastructure, and devices in systems are growing higher heterogeneous and complex. In order to practically optimize, manage, organize, maintain systems, more intelligence needs to be used. Therefore, it became necessary to utilize ML methods [17].

ML is utilized in several fields of computer designing and programing algorithms with high achievement, for instance, detect face & shape, filtering email spam, medical determination, predicting traffic, recognizing character, the Google cars self-driving, Netflix, recommendation engines online [18].

The ML is either supervised or unsupervised. Also, it has several models; each model differs from the other in terms of speed and accuracy. There are several types of ML such as Naïve Bayes, ensemble learning, Decision Trees, artificial neural networks, support vector machines, and convolutional neural networks. The models employed in this study will be explained in detail as follows:

3.1   Naïve Bayes

Naive Bayes is a supervised ML techniques classification. It is a type of simplistic probabilistic classification technique based on implementing Bayes' theorem with the presumption of independence between features. It is a simple method for constructing classifiers: models that select class labels to problem cases served as vectors of feature values, where the class labels are represented from some restricted set. It is also useful for multi-dimensional data as the likelihood of each feature is estimated independently. There are three standard NB classifiers; these are Bernoulli NB classifier, Gaussian NB Classifier, and Multinomial NB classifier [19]. Bayesian classification is based on Bayes Theorem, and Bayes Theorem is declared as Eq. (1) [21]:

$$P(D/Y) = \frac{P(Y/D).P(D)}{P(Y)} \qquad (1)$$

where

P(D|Y) is the next probability of the target class.
P(D) is called the previous probability of class.
P(Y|D) is the likelihood which is the probability of predictor of assigned class.
P(Y) is the previous probability of predictor of class.

The NB classifier operates as follows:

- Let Z be the training dataset associated with class labels. The N-dimensional element represents each tuple vector, Y = (y1, y2, y3,..,yn).
- Consider that there are n classes D1, D2, D3...., Dm. The classifier will be predicted that Y relates to the class with a higher next probability, adapted on Y. The NB classifier indicates an unknown tuple Y to the class Di if and only if P(Di| Y) > P(Dj|Y) For $1 \leq j \leq m$, and $i \neq j$, above following probabilities are estimated using Bayes Theorem.

The advantage of the NB is it does not requires high computational time for training [20]. It increases the classification achievement by eliminating unnecessary features. The disadvantages of the NB are it requires huge records to achieve excellent results. Less accurate as contrasted to other classifiers on the same data.

3.2 Decision Trees

Decision Trees are classification or regression models expressed in a tree-like structure [21]. It is a supervised learning method and intended to classify or predict a discrete division from the data. In the ML sense, the purpose is to design a classification model that predicts a target class by learning decision rules induced from the data features [22]. In Fig. 2, an internal node N indicates a test on an attribute; an edge E describes an outcome of the test. The Leaf means C depicts class labels or class distribution. Four reasons excited to work with decision tree classifiers:

1. It can be learned and updated from data relatively fast compared to another technique.
2. It is more intuitive, more comfortable, and simpler to interpret and assimilate by humans.
3. Unlike another classification technique, with DT classifiers, one can perform data-driven root cause analysis of faults; one can follow a path from the end state to the initiating event, a way that follows the sequence of how events are interlinked.
4. It has a high accuracy compared to other techniques and demonstrates the best combination of speed and error rate.

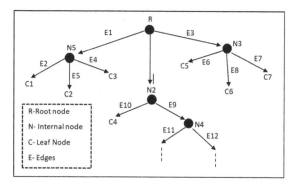

Fig. 2. Graphical representation of a Decision Tree classifier [21].

From Fig. 2, an internal node N indicates a test on an attribute, an edge E depicts an outcome of the test, and the Leaf nodes C depict class labels or class distribution.

3.3 Support Vector Machines (SVMs)

SVMs are a kind of supervised ML technique. In a two-class classification problem, the principal object is to construct a model that assigns each new sample in a suitable class. SVMs techniques attempt to determine this task by exerting the training samples into a more distinguished dimension where they are linearly separate and allocated to a class. The linearly separable binary class datasets are simple to define since the decision boundary of the two groups is just a straight line [23]. Figure 3 illustrates the separate features into two regions [23].

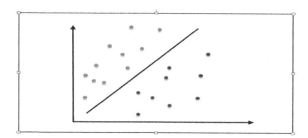

Fig. 3. Separate the features into two regions [23].

The aim is to use the original feature space situations where they are non-linearly separating to a new feature space. A hyperplane is formed in this new space as a straight line in two dimensions. It acts as a boundary of decision that separates the data; the maximum margin hyperplane is associated with this boundary. The training locations that are nearest to the decision boundary are named support vectors. It can define the maximum-margin hyperplane for the learning problem. In this style, SVM explores a maximum margin hyperplane to separate the data with the samples on the border named support vectors. Figure 4 illustrates the structure of the linear SVM [24].

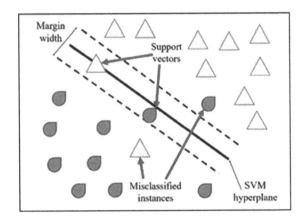

Fig. 4. Structure of the linear SVM [24].

Each new entry will be used in this new space, wherever it will be classified depending on the region.

In SVM, the optimal hyperplane is bound to maximize the capability of the model. But, if the training data are not separated linearly, the classifier obtained may not have a high generalization ability, even if the hyperplanes are optimally bound. To solve this problem, the kernel function is used. Some of the most used kernel functions are [25]:

1. Linear kernel.
2. Polynomial kernel.
3. Radial Basis Function (RBF) kernel.
4. Sigmoid kernel.

RBF kernels are the common generalized method of kernelization and are one of the commonly widely utilized kernels due to their relation to the Gaussian distribution. The RBF kernel function for two points xi and xj calculates the identity or how close they are to each other. This kernel can be mathematically expressed as a formula in Eq. (2) [25, 26]:

$$K(xi, xj) = e^{\frac{(Xi-Xj)^2}{2\sigma^2}} \tag{2}$$

where σ is the width of Gaussian kernel.

4 Fall Detection System

In this study, an intelligent system is suggested to detect the fall and reduce the false alarm effectively. The utilizing data are extracted from Kinect cameras, and s ADXL345 accelerometer sensor are installed in an elderly environment to monitor all activities. The implementation process of the proposed research is employed three types of ML techniques. These techniques are SVM, NB, and DT.

4.1 Data Source

The proposed system utilized publicly available datasets URFD [27]. It contains 11 features; each feature includes 11544 records. URFD includes sequences of 40 ADLs and 30 falls. Fall events are captured with two Kinects cameras (Camera 0 and Camera 1), corresponding with the accelerometer sensor. ADL events are captured with only one Kinects camera (camera 0) corresponding with the accelerometer sensor. Each event includes a sequence of RGB images, depth images, and accelerometer data. In each scenery, various people do the same actions. Two types of falls were done by five volunteers and recognized by two overlapped views: sitting to falling and standing. The URFD dataset is constructed as follows: Every row includes a sequence of RGB and depth images for camera 1 and camera 0(installed on the ceiling and floor, respectively), synchronization data, and raw accelerometer data. Every video is deposited in a separate zip file and archived in Portable Network Graphics (PNG) image sequence. The depth image is deposited in PNG16 format. Data synchronization is deposited in Comma Separated Values (CSV) format and contains frame number, time in milliseconds since sequence start, and interpolated accelerometer data corresponding to the image frame. Raw accelerometer data is deposited in CSV format and contains time in milliseconds since sequence start and accelerometer data: ACC total, Ax, Ay, Az. All accelerometer data are in gravity units (g). Extracted features from depth maps are deposited in CSV format. Each row contains one sample of data corresponding to one depth image. Figure 3.1 depicted the URFD dataset. The following features are organized as follows:

- sequence name: camera name is omitted because all of the samples are from the front camera ('fall-01-cam0-d' is 'fall-01', 'adl-01-cam0-d' is 'adl-01' and so on).
- Frame number: corresponding to the number in a sequence.
- HeightWidthRatio: bounding box height to width ratio.
- MajorMinorRatio: ratio of major to the minor axis, calculated by Binary Large Object Both (BLOB) technique of segmented person.
- Bounding_Box_Occupancy: ratio of how the bounding box is filled by the pixels of the individual.
- MaxStdXZ: is the standard deviation of pixels from the center of the X-axis and the Z-axis, respectively.
- HHmaxRatio: rate of the height of a person in the frame to the actual height.
- H: actual person height measured in millimeters.
- D: distance of the center of person to the floor measured in millimeter.
- P40: ratio of the number of the point clouds relating to the cuboid of 40 cm height and located on the floor to the number of the point clouds relating to the cuboid of measurement equal to person's height.
- Label: describes human posture in the depth frame; ' -1 ' means the person is not lying, '1' means the person is lying on the ground; '0' is a temporary pose, when the person "is falling", the frames '0' are not used in classification. Figure 5 illustrates The URFD dataset.

	A	B	C	D	E	F	G	H	I	J	K
1	sequence_name	frame No	HWRatio	MMRatio	BOccupanc	MaxStdXZ	HHRatio	H	D	P40 - ratio	label
2	fall-01	1	3.1667	2.9098	0.55367	126.026	1.0324	1899.54	1056	0.04731	-1
3	fall-01	2	3.3067	2.9699	0.47876	125.566	1.1251	2070.12	1065.95	0.04818	-1
4	fall-01	3	3.1408	3.0506	0.54374	123.157	1.0161	1869.64	1055.5	0.05018	-1
5	fall-01	4	3.4306	3.1435	0.48859	124.561	1.1251	2070.12	1076.15	0.04788	-1
6	fall-01	5	3.6324	3.3012	0.49744	123.609	1.1251	2070.12	1075.51	0.05254	-1
7	fall-01	6	3.3788	3.4746	0.54308	121.883	1.0174	1872.02	1065.9	0.05492	-1
8	fall-01	7	3.3636	3.5179	0.52355	122.429	0.98797	1817.86	1076.02	0.05958	-1
9	fall-01	8	3.1739	3.2962	0.48442	123.825	0.97907	1801.49	1074.77	0.04781	-1
10	fall-01	9	3.4531	3.4923	0.50834	123.525	0.97907	1801.49	1067.87	0.05104	-1
11	fall-01	10	3.6618	3.4449	0.4371	127.327	1.1251	2070.12	1090.54	0.04364	-1
12	fall-01	11	3.2424	3.5455	0.46701	124.823	0.98699	1816.07	1100.87	0.03699	-1
13	fall-01	12	3.4925	3.783	0.42289	123.076	1.0324	1899.54	1081.1	0.05475	-1
14	fall-01	13	3.0882	3.4521	0.44307	123.166	0.97963	1802.52	1087.91	0.03019	-1
15	fall-01	14	2.6515	3.0049	0.48407	115.935	0.85824	1579.16	1014.24	0.03202	-1

Fig. 5. The URFD dataset

4.2   Methodology

Firstly, it is important to discriminate and understand the fundamental differences between falls and ADL to implement the FD system successfully. The proposed methodology consists of several stages. The methodology and process will be explained in detail as follow:

Feature Utilization
The proposed approach utilizes features; these features are extracted from the Kinect camera and accelerometer sensor. The initial features are extracted from the Kinects camera. The other features are extracted from the ADXL345 accelerometer by calculating the total acceleration. The total acceleration is calculated by applying the following formula in Eq. 3.

$$Total\_ACC = \sqrt{x^2 + y^2 + z^2} \qquad (3)$$

Preprocessing
The proposed system consists of two pre-processing operations, these are feature encoding and Scaling. Feature encoding is necessary because the ML typically deals with numeric data. There are several techniques that achieve this operation; one of these techniques is using "case" instruction. However, this method is not practical because when adding new data, it is done manually. The two most popular method to perform this operation are using OneHotEncoder or LabelEncoder. LabelEncoder encodes labels with a value within 0 and m-1. If a label repeats, it assigns the same value as previously assigned. The system employs Python LabelEncoder instruction to perform this operation. For example, the 'fall-01' and 'adl-01' in the column 'sequence_name' is encoded to '40' and '0', respectively. Feature Scaling is a procedure that standardizes the features being in the data in a fixed range. Since the data used in this system contains large numbers such as 2165.38 in feature 'D' and contains small numbers such as 0.102 in the feature 'HHRatio', standardization is a very productive

process re-scales a feature value. It can calculate by using the following formula in Eq. (4) [28]:

$$X_{stand} = \frac{x - mean(x)}{Standard\ deviation} \tag{4}$$

The block diagram of the proposed FD system illustrates in Fig. 6.

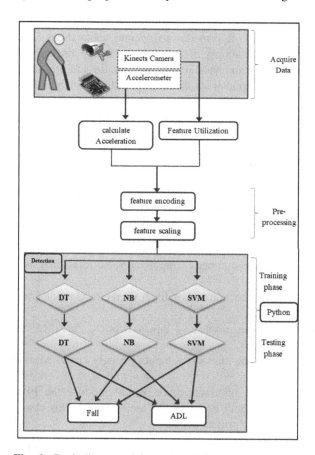

Fig. 6. Bock diagram of the proposed fall detection system

Figure 6 explains the life cycle of the proposed system. This cycle begins with the dataset's acquisition, followed by extracting the features and then pre-processing them. Through the SVM, NB, and DT techniques, applying the training and testing process, and obtaining the results.

4.3   Programming Environment

This system was implemented with Spyder. It is a scientific environment for python 3.8 development and includs with Anaconda. It is robust interactive with precedent editing, interactive testing, introspection, and debugging features.

Anaconda is an available and open-source, straightforward to establish the distribution of Python programming languages. It presents an effective environment accepted for statistical analysis, data science, ML, and scientific computing.

Anaconda Navigator is a desktop Graphical User Interface (GUI) incorporated in the Anaconda distribution. It provides to operate applications implemented in the Anaconda distribution and simply manages Anaconda channels, packages, and environments without using command-line instructions. In addition to the Spyder, Anaconda includes a set of applications that can be utilized, such as JupyterLab, Jupyter Notebook, Qt Console, Glueviz, Orange3, and Visual Studio Code. One of the characteristics that distinguish the Anaconda is the ease of adding libraries to it.

5   Evaluation Metrics

This section includes the method, the metrics, and the premises, that are utilized to evaluate the performance of the suggested system. The evaluation of the classification technique depends on the following metrics: accuracy, precision, recall and executed time. The calculation of these measures depends on the following descriptions [11]:

- True Positive (TP): the number of inputs correctly classified as positives, ADLs classified as ADLs.
- True Negative (TN): the number of input correctly classified as negatives, Falls classified as Falls.
- False Positive (FP): the number of inputs wrongly classified as positives, Falls classified as ADLs.
- False Negative (FN): the number of inputs wrongly classified as negatives, ADLs classified as Falls.
- The ideal performance of classification would produce the FP= 0 and FN= 0. However, in exercise, both FNs and FPs are non-zero. This case negatively influences the algorithm performance.

The three metrics can be more well-reviewed applying the following formulations [11]:

- Accuracy (AC): the ratio between the actual Falls and ADLs over all the inputs activities. It can calculate by using the following formula in Eq. 5:

$$AC = \frac{(TP + TN)}{(TP + TN + FP + FN)} \tag{5}$$

- Precision (PR): the ratio between the actual ADLs over all the inputs classified as ADLs. It can calculate by using the following formula in Eq. 6:

$$PR = \frac{TP}{TP + FP} \tag{6}$$

– Recall (RE): the ratio between the number of input classified as ADLs over all the actual ADLs. It can calculate by using the following formula in Eq. 7:

$$RE = \frac{TP}{TP + FN} \tag{7}$$

6   Experimental Result

In this section, the proposed system's role is explained in developing and enhancing current detection systems. The performance metrics are also calculated, which are the confusing matrix, accuracy, precision, recall, and executed time for the proposed system method. Three types of machine learning techniques will be used in the proposed system: SVM, NB, and DT. By applying SVM to a set of features, these features are extracted from the Kinect camera and accelerometer sensor. Since kernel selecting is one of the essential issues in the training step of SVM, Therefore, more than one kernel can be used, such as linear, Polynomial, Gaussian (RBF), and Sigmoid techniques. The results are obtained through trial and error shown in Table 1 and Fig. 7.

Table 1. SVM kernels

Kernels	Linear	Polynomial	Sigmoid	RBF
Accuracy	99.83	99.22	93.72	99.93

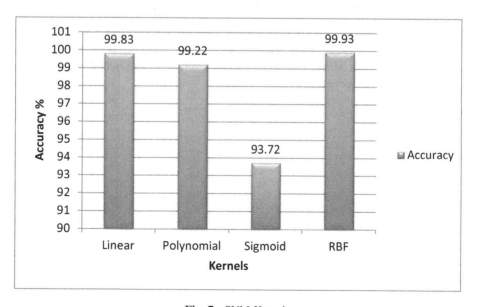

Fig. 7. SVM Kernels

According to the results in Table 1 and Fig. 7, the highest accuracy ratio was obtained from using the 'RBF' Kernel; therefore, the proposed system will be used this Kernel. The performance metrics are shown in Table 2 and Fig. 8.

Table 2. SVM performance matrix

Accuracy	False alarm	Precision	Recall	Time
99.93	0.07	99.95	99.95	0.354

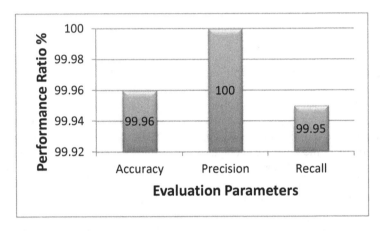

Fig. 8. SVM performance

According to the results in Table 3 and Fig. 8, FD system that based on SVM algorithm obtained an accuracy, precision, and recall ratio reached to 99.96, 100 and 99.95, respectively.

NB is applying to the same features that applied in SVM. Since there are three common models of NB classifiers, are Bernoulli NB classifier, Gaussian NB Classifier, and Multinomial NB classifier, through trial and error the accuracy that is obtained by applying these three models are shown in Fig. 9.

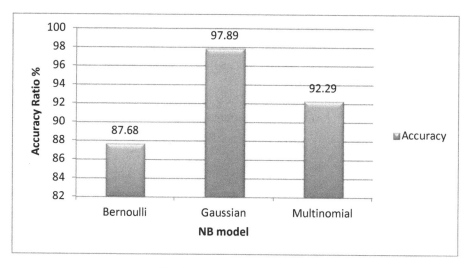

Fig. 9. Accuracy of NB models

According to the results in Fig. 9, the accuracy obtained by utilized the Bernoulli NB, Gaussian NB, and Multinomial NB is 88.94, 98.1, and 91.72, respectively. Since Gaussian NB obtained the highest accuracy, it will be approved in the proposed system. The performance metrics showed in Table 3 and Fig. 10.

Table 3. NB performance matrix

Accuracy	False alarm	Precision	Recall	Time
97.89	2.11	97.22	100	0.168

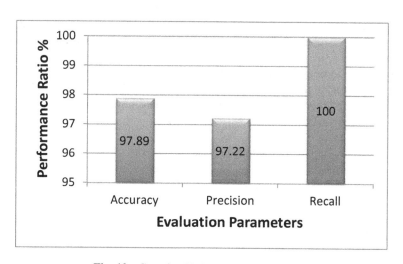

Fig. 10. Gaussian Naïve Bayes performance

According to the results in Table 3 and Fig. 10, FD system based on SVM algorithm obtained the following results 97.89, 97.22, and 100 for accuracy, precision, and recall ratio respectively.

DT is applying to the same features that applied in SVM and NB. The performance metrics showed in Table 4 and Fig. 11.

Table 4. DT performance matrix

Accuracy	False alarm	Precision	Recall	Time
99.96	0.04	100	99.95	0.771

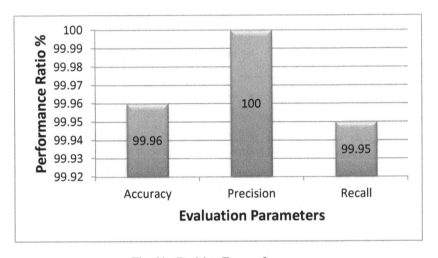

Fig. 11. Decision Tree performance

According to the results in Table 4 and Fig. 11, FD system that based on DT algorithm, they obtained 99.96, 100 and 99.95 accuracy, precision, and recall ratio, respectively.

Complexity, in one of its forms, refers to the relationship between the increase in the time it takes an algorithm or program to run, and the increase in the input size of that algorithm or program. Table 5 show the standard complexity of each technique used in proposed study.

Table 5. Standard complexity

Technique	Complexity
SVM	$O(N^3)$
NB	$O(N^d)$
DT	$O(N \, Log(d))$

Machine Learning models have something particular to them: the complexity is not only a function of the input size N, but also of the number of features of our data d.

7   Comparison of Classification Techniques

The intention of this study is to investigate the performance of different classification techniques for a set of data. The tested techniques or algorithms are Support Vector Machine, Naive Bayes, and Decision Tree. Table 6 and Fig. 12 presented the performance comparison with existing works of these techniques.

Table 6. Performance comparison with existing works

Technique	Accuracy	Precision	Recall	Time
SVM	99.96	100	99.95	0.272
NB	98.1	97.48	100	0.168
DT	99.96	100	99.95	0.771

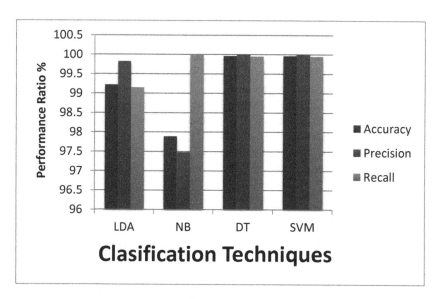

Fig. 12. Performance comparison with existing works

Based on Fig. 12 and Table 6, we can clearly recognize that the highest accuracy is 99.96%, and the lowest is 98.1%. In fact, the highest accuracy relates to the SVM classifier and DT classifier, and the most insufficient accuracy refers to the NB classifier. Figure 13 presented the time compared with existing works of these techniques.

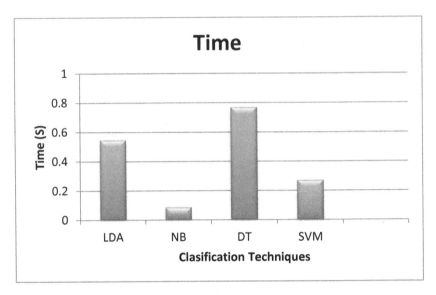

Fig. 13. Time comparison with existing works

Figure 13 showed the time in seconds for each technique; the fast algorithm is NB with 0.168 s execution time, followed by the SVM algorithm, and the slowest algorithm is DT with 0.771 s execution time.

Table 6 and Figs. 12 and 13 indicated that the optimum accuracy was obtained by applying the SVM and DT techniques. By comparing the execution time between these two methods, it was clear that the SVM was the fastest. Hence, SVM was best technique utilized the URFD dataset to discriminate fall from ADL.

8   Conclusion

Fall detection is one of the crucial mechanisms that can enable better care particularly for the elderly people who are more prone to fall, to get emergency assistance. In this paper, we have presented an intelligent system based on ML techniques. The experimental results demonstrate that the proposed fall detection scheme has achieved higher performance compared to methods on challenging URFD dataset. The main motivation for the FD approach is to preserve the lives of people, especially the elderly, because they are vulnerable to falling at any time. Hence, health expenses will be reduced. The proposed FD approach has some limitations: Time is the major challenge, and this study has conducted experiments on a relatively non-real-time dataset of FD. Exploring real-time reviews would contribute towards discovering new side-effects of activities.

In the future, we will serve to improve the approach functionality and provide an additional evaluation of its performance by using different features. Therefore, future research is to discover more features. Further work may focus on online monitoring and messaging system for mobile phones.

References

1. Howedi, A., Lotfi, A., Pourabdollah, A.: Accelerometer-based human fall detection using fuzzy entropy. In: IEEE International Conference on Fuzzy Systems, 2020-July (2020)
2. Martínez-Villaseñor, L., Ponce, H., Brieva, J., Moya-Albor, E., Núñez-Martínez, J., Peñafort-Asturiano, C.: Up-fall detection dataset: a multimodal approach. Sensors (Switzerland) **19** (2019)
3. Islam, M.M., et al.: Deep learning based systems developed for fall detection: a review. IEEE Access **8**, 166117–166137 (2020)
4. Ismail, M.M., Ben, A.A., Bchir, O.: Fall detection using the histogram of oriented gradients and decision-based fusion. J. Comput. Sci. **16**, 257–265 (2020)
5. Liakos, K.G., Busato, P., Moshou, D., Pearson, S., Bochtis, D.: Machine learning in agriculture: a review. Sensors (Switzerland) **18**, 1–29 (2018)
6. De Quadros, T., Lazzaretti, A.E., Schneider, F.K.: A movement decomposition and machine learning-based fall detection system using wrist wearable device. IEEE Sens. J. **18**, 5082–5089 (2018)
7. Alves, J., Silva, J., Grifo, E., Resende, C., Sousa, I.: Wearable embedded intelligence for detection of falls independently of on-body location. Sensors (Switzerland) **19** (2019)
8. Nooruddin, S., Milon Islam, M., Sharna, F.A.: An IoT based device-type invariant fall detection system. Internet of Things **9**, 100130 (2020)
9. Jeffin Gracewell, J., Pavalarajan, S.: Fall detection based on posture classification for smart home environment. J. Ambient Intell. Humaniz. Comput. (2019)
10. Xi, X., Jiang, W., Lü, Z., Miran, S.M., Luo, Z.Z.: Daily activity monitoring and fall detection based on surface electromyography and plantar pressure. Complexity **2020** (2020)
11. Elshwemy, F.A., Elbasiony, R., Saidahmed, M.T.: An enhanced fall detection approach in smart homes using optical flow and residual autoencoder. Int. J. Adv. Trends Comput. Sci. Eng. **9**, 3624–3631 (2020)
12. Jansi, R., Amutha, R.: Detection of fall for the elderly in an indoor environment using a tri-axial accelerometer and Kinect depth data. Multidimension. Syst. Signal Process. **31**(4), 1207–1225 (2020). https://doi.org/10.1007/s11045-020-00705-4
13. Wang, B.H., Yu, J., Wang, K., Bao, X.Y., Mao, K.M.: Fall detection based on dual-channel feature integration. IEEE Access **8**, 103443–103453 (2020)
14. Ogawa, Y., Naito, K.: Fall detection scheme based on temperature distribution with IR array sensor. In: Digest of Technical Papers - IEEE International Conference on Consumer Electronics, 2020-January, pp. 1–5 (2020)
15. Alarifi, A., Alwadain, A.: Killer heuristic optimized convolution neural network-based fall detection with wearable IoT sensor devices. Meas.: J. Int. Meas. Confed. **167**, 108258 (2020)
16. Tahir, A., Ahmad, J., Morison, G., Larijani, H., Gibson, R.M., Skelton, D.A.: HRNN4F: hybrid deep random neural network for multi-channel fall activity detection. Probab. Eng. Inf. Sci. **35**, 37–50 (2021)
17. Xie, J., Richard Yu, F., Huang, T., Xie, R., Liu, J., Wang, C., Liu, Y.: A survey of machine learning techniques applied to software defined networking (SDN): research issues and challenges. IEEE Commun. Surv. Tutor. **21**, 393–430 (2019)
18. Alzubi, J., Nayyar, A., Kumar, A.: Machine learning from theory to algorithms: an overview. J. Phys.: Conf. Ser. **1142**, 15 (2018)
19. Singh, G., Kumar, B., Gaur, L., Tyagi, A.: Comparison between multinomial and Bernoulli Naïve Bayes for text classification. In: 2019 International Conference on Automation, Computational and Technology Management, ICACTM 2019, pp. 593–596 (2019)

20. Jadhav, S.D., Channe, H.P.: Comparative study of K-NN, Naive Bayes and Decision Tree Classification techniques. Int. J. Sci. Res. (IJSR) **5**, 1842–1845 (2016)
21. Abdallah, I., et al.: Fault diagnosis of wind turbine structures using decision tree learning algorithms with big data. In: Safety and Reliability - Safe Societies in a Changing World - Proceedings of the 28th International European Safety and Reliability Conference, ESREL 2018, pp. 3053–3062 (2018)
22. Purdilă, V., Pentiuc, ŞG.: Fast decision tree algorithm. Adv. Electr. Comput. Eng. **14**, 65–68 (2014)
23. Hossain, S.F.: An Automated Directive Fall Detection System Using Single 3D Accelerometer and Learning Classifier (2017)
24. Kepski, M., Kwolek, B.: Fall detection on embedded platform using Kinect and wireless accelerometer. In: Miesenberger, K., Karshmer, A., Penaz, P., Zagler, W. (eds.) ICCHP 2012. LNCS, vol. 7383, pp. 407–414. Springer, Heidelberg (2012). https://doi.org/10.1007/978-3-642-31534-3_60
25. Cervantes, J., Garcia-Lamont, F., Rodríguez-Mazahua, L., Lopez, A.: A comprehensive survey on support vector machine classification: applications, challenges and trends. Neurocomputing (2020)
26. Zerrouki, N., Harrou, F., Sun, Y., Houacine, A.: Vision-based human action classification using adaptive boosting algorithm. IEEE Sens. J. **18**, 5115–5121 (2018)
27. Khraief, C., Benzarti, F., Amiri, H.: Elderly fall detection based on multi-stream deep convolutional networks. Multimed. Tools Appl. **79**(27–28), 19537–19560 (2020). https://doi.org/10.1007/s11042-020-08812-x
28. Wang, D., Wang, X., Zhang, Y., Jin, L.: Detection of power grid disturbances and cyber-attacks based on machine learning. J. Inf. Secur. Appl. **46**, 42–52 (2019)

Three N-grams Based Language Model for Auto-correction of Speech Recognition Errors

Imad Qasim Habeeb[1]([✉]) [iD], Hanan Najm Abdulkhudhur[2] [iD],
and Zeyad Qasim Al-Zaydi[3] [iD]

[1] College of Medical Informatics, University of Information Technology
and Communications, Baghdad, Iraq
emadkassam@uoitc.edu.iq
[2] Directorate of Education Al-Karkh 2, Ministry of Education, Baghdad, Iraq
[3] Biomedical Engineering, University of Technology, Baghdad, Iraq
11344@uotechnology.edu.iq

Abstract. Some applications of Automatic Speech Recognition (ASR) can cause errors, especially when an environment of speech contains noise. This is one of the major barriers to the widespread adoption of speech recognition, especially for applications with a large dictionary. The persistence of ASR errors has increased the need for new techniques to automatically correct these errors. In this paper, a novel three n-grams based language model that uses Google n-gram files was designed. The proposed method performed auto-detection and correction of errors in ASR output text. It consists of two parts. The first part explains the steps for designing and building a language model database using data from Google n-gram files, while the second part shows how to use the context information of the first part in ASR troubleshooting. In addition, this method includes multi-pass filtering which leads to faster processing time and improved computation efficiency. Each pass acts as an error filter. Experimental results showed the success of using this method in reducing the rate of speech errors in the ASR text. It achieved a 15.71% relative decrease against the best Word Error Rate for the comparative methods.

Keywords: ASR errors · Auto-correction · ASR post-processing · Language model · N-grams

1 Introduction

Automatic Speech Recognition (ASR) is used to extract series of words from human speech [1]. Output text of ASR almost always produces errors, especially when the environment of speech contains noise [2–4]. The proposed method performs multi-pass filtering in detecting and correcting these errors. In addition, it combines three N-grams-based language models with the *Levenshtein* distance algorithm [5] for auto-detection and correction of errors in ASR output text. N-gram language model is a probabilistic model with two functions. The first function is the ability to predict the next candidate from the previous $(k-1)$ words while the second is to predict the

© Springer Nature Switzerland AG 2021
A. M. Al-Bakry et al. (Eds.): NTICT 2021, CCIS 1511, pp. 131–143, 2021.
https://doi.org/10.1007/978-3-030-93417-0_9

probability of the whole phrase [6]. Both functions have a drawback when used alone in ASR post-processing error correction. The ASR post-processing is a process implemented after the text of ASR is produced to correct its output errors. The first function drawback is that it uses previous $(k-1)$ words only to predict the next candidate w_i w_i in the phrase. Hence, it ignores the influence of next $(n+1)$ words to predict the suitable candidate for incorrect words.

For example, the phrase "*the teacher tried to "expilian" the problem*" has a single misspelled word "*expilian*". Therefore, the list of candidates with high frequency that obtained from the *Google* n-gram website for auto-correction is: "make", "explain", and "train". The Language model may replace the misspelled word "*expilian*" by the candidate "*make*" due to its appearance is more frequent than two other candidates in *Google* n-gram. This indicates that used only previous $(k-1)$ words to predict the next candidate is not enough to correct misspelled words because it ignores the influence of next $(n+1)$ words ("*the problem*") in the phrase that comes after the wrong word "*expilian*".

The second function drawback of the language model is that it uses the probability of the whole phrase to predict its validity. For example, the phrase "*I traveled to Sbain in the summer*" has also the single misspelled word "*Sbain*". Therefore, the list of candidate phrases that resulted from the *Google* n-gram website is: "China", "France", "Spain", and "Mexico". The language model may also replace the phrase that contains "*Sbain*" with the phrase that contains "*France*" duo to the candidate "*France*" is more frequent than other Candidates. Both examples present the importance of using novel methods to overcome the drawbacks of the Language model.

This research was organized into five section [7]: the next section describes related work on ASR post-processing error correction; Sect. 3 explains the steps for the proposed method design. In Sect. 4, the implantation, data collected, experiments, and evaluation are shown. The last section includes research conclusions and future work of our study.

2   Related Work

Several methods are used to detect and correct ASR errors, such as context-based methods and rules-based methods. In that context, Kaki, Sumita, and Iida [8] presented a method that consists of two processes. The first process used erroneous-correct utterance pairs to detect and correct most errors. The second process corrects the remaining errors according to the best similarity between the incorrect word and candidates. Experimental results on the Japanese corpus show reduction of 8.5% in ASR errors.

Mangu and Padmanabhan [9] suggested a confusion network that can learn rules in training, these rules can learn the sequence of word confusions and learn the previous probability of each word, then their method used these rules at testing in selecting the best candidate to replace with the incorrect word. Experimental results on the corpus of Switchboard show an improvement of 10% in word error rate if the errors in confusion network and less if outside of it. Jung, Jeong, and Lee [10] designed a syllable-based noisy channel model to deal with syllables instead of words, this model learns from transformation between syllables. Experimental results on the Korea corpus show improvement of 40% in the error correction rate. Zhou, Meng, and Lo [11] present a multi-pass framework to detect and correct the errors in ASR by combining a model of mutual information and a fixed trigram language model, Experimental results show a relative reduction of 4% in character error rate (CER).

Ringger and Allen [12] proposed a post-processing framework, that consists of a noisy channel model for detecting errors based on patterns and a fixed bigram language model. Their framework selects the best candidate for incorrect words. Experimental results on the English corpus show an improvement of 20% in the error correction rate. The last one is Habeeb, Fadhil [13] who proposed an ensemble method that merges three noise reduction filters to improve the ASR output. Their method does not rely on one filter for perfect noise reduction but incorporates information from three filters. After that, n-copies of the speech signal were generated as the major factor for their technique. The different features of these n-copies can be used to extract various texts from ASR system. Therefore, the best among these outputs can be elected as the result of ASR output. Their method was compared with three related several methods. The experimental tests displayed a relative decrease by 16.61% on error rate. However, their method is only suitable for the recognition of a speech signal in the presence of noise.

3   Proposed Technique

The proposed method used three n-grams. Their names are middle n-gram, left n-gram, and right n-gram. Also, the proposed method consists of multi-pass filtering. The first filtering pass, second filtering pass, third filtering pass, and fourth filtering pass are middle n-gram, *Levenshtein* distance algorithm, left n-gram, and right n-gram respectively. Each filtering pass tries to find one candidate for an incorrect word, if exists, then no need for to next filtering pass, otherwise, if several candidates are produced, the next filtering pass is performed, and so on until the last one. The purpose of multi-pass filtering is that most errors can be corrected with first or second filtering, thus no need for more calculation and thus reduces processing time. On the other hand, complex errors need extra calculation to correct them and thus need other filtering passes. The methodology used in this research to detect and correct ASR errors consists of three stages: (1) designing and implementing a language model database using the data from *Google* n-gram database, (2) detecting the errors in output ASR text, (3) correcting these errors in output ASR text based on multi-pass filtering.

3.1   Language Model Design

Any strong language model requires a large-scale corpus to train and to provide probability correctly. This study used Google n-gram files [14], which contain 1 to 5 g and their frequency. Google collects the words and phrases of these files from Web servers. Google n-gram files are available free and can be downloaded in the txt format. The selecting of Google n-gram in this research due to two reasons. First, it is a large free source of web corpora. Second, it contains more than 100,119,665,584 phrase in various categories. The last, the data can be useful for the speech recognition and other applications [15, 16].

All *Google* n-gram files were stored in five tables in the database because dealing with the database is better than through multi-text files. Microsoft SQL Server 2014 is used to design a database and to save the data. Table 1 shows a data model, which is used for all tables in the database with a unique name for each one. Names of tables are ("*gram1*", "*gram2*", "*gram3*", "*gram4*", and "*gram5*") and these tables consists of two columns each.

Table 1. Language model database.

Column name	Data type	Description
word_or_sentence	nvarchar	Word or sentence (Primary key)
Frequency	Integer	Frequency of the word or sentence

Figure 1 illustrates the process of transferring and processing the data extracted from Google n-gram files and storing it in the language model database. In the beginning, all *Google* n-gram files are stored in one folder, and then, the program processes the files one by one in sequence. In the processing stage, all n-grams are transformed into the lower case to reduce the number of n-grams and this creates many duplicate n-grams, all duplicates are merged by adding their frequencies. Each sentence or word from *Google* n-gram was stored in field "*word_or_sentence*" while their frequency was stored in field "*frequency*" in the corresponding table. The process has been continued until all *Google* n-gram files are handled.

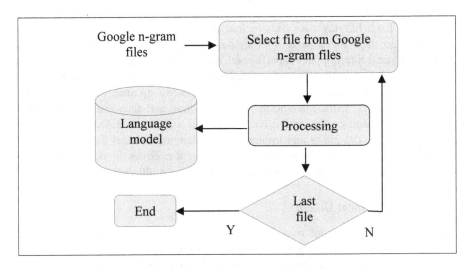

Fig. 1. Filling of language model database.

3.2   Detecting of ASR Text Errors

Initially, non-word errors are easily identified through the table *"gram1"*, in which any word outside of the *"gram1"* is considered a non-word. The process of detecting non-word errors in output ASR text is shown in Fig. 2.

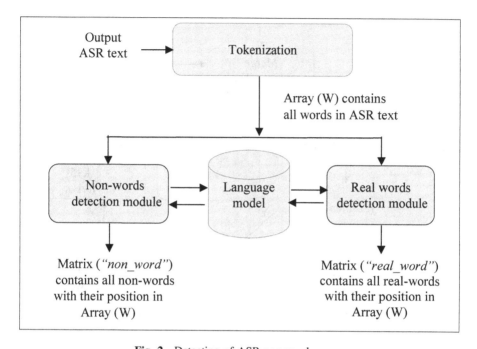

Fig. 2. Detecting of ASR non-word errors.

In the first stage, which is called tokenization, the output ASR text is divided into words using the space as a divider between words, then they were stored in an array called "W". In the non-word detection module, array "W" is converted to a table called *"all_word"* so that it can send SQL query to the database to filter both tables *"all_-word"* and *"gram1"*, where any word from table *"all_word"* does not exist in the table *"gram1"* is considered a non-word. The result of the SQL query is a table contains all words that do not exist in *"gram1"*, this table is converted to a matrix called *"non_-word"* to speed up operations on it. The matrix *"non_word"* contains all non-words with their position in array "W".

Identifying real-word errors are more complex than non-word errors because they already exist in the table *"gram1"*. The problem with them is that their words are not suitable for the sentence. Hence, the real-words detection module will process array "W" to produce a matrix called *"real_word"*, which contains all real-words with their position in array "W". Figure 3 shows the steps in the real-words detection module. The steps are:

Step 1: merge words $W[i]$, $W[i+1]$, $W[i+2]$, $W[i+3]$ and $W[i+4]$ in one sentence using space among them. Note that the initial value of "i" is equal to one. The *check module* is used to check the existence of this sentence in the corresponding n-gram table in the database.

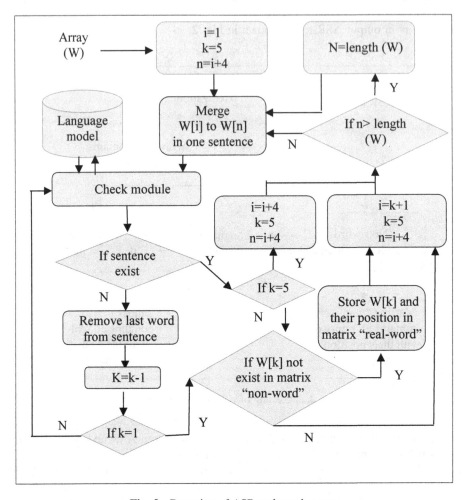

Fig. 3. Detecting of ASR real-word errors.

In case of all sentences exist in table "*gram5*" then they consider as free from real-word error. Also, in this case, the value of "*i*" will increase by four so that will become equal to ("*i = i+4*") and so on until the end of the length of array "W", on the other hand, "*i*" increased by four, not five because maybe both previous sentence and next sentence have existed in table "*gram5*", but there is no relation between them as shown in Fig. 4.

Step 2: in case of any sentence is not exist in table "*gram5*", then step 2 will start by removing the last word W[*i*+4] from the sentence, and check if the sentence exists in table "*gram4*" if it exists then the word W[*i*+4] is considered as real word error and value of "*i*" will become equal to "*i = i+5*". If else and it not exists then step 3 will start.

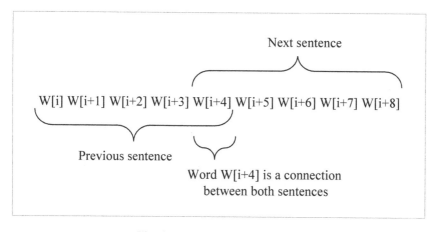

Fig. 4. Sentence by sentence checking.

Step 3: will start by removing the last word W[i+3] from the sentence, and check if the sentence exists in table "*gram3*" if it exists then the word W[i+3] is considered as real word error and value of "i" will become equal to "$i = i+4$". If else and it not exists then step 4 will start.

Step 4 will also start by removing the last word W[i+2] from the sentence, and check if the sentence exists in table "*gram2*" if it exists then the word W[i+2] is considered as real word error and value of "i" will become equal to "$i = i+3$". If else and it not exists then step 5 will start.

Step 5 will start by removing the last word W[i+1] from the sentence, and check if the sentence exists in table "*gram1*" if it exists then the word W[i+1] is considered as real word error and value of "i" will become equal to "$i = i+2$". If else and it not exists then step 6 will start.

Step 6 will consider the word W[i] as a real word error and the value of "i" will become equal to "$i = i+1$".

3.3   Correcting of ASR Text Errors

Figure 5 shows the process of correcting the errors in output ASR text, Matrixes "*real_word*" & "*non_word*" that store all errors with their position in array "W" is passed to the correction process. All incorrect words will move one by one in sequence to the correction module, which is used to select the best candidate for each incorrect word and to replace it with this candidate. Our language model has been used to give the probability for each candidate.

This research used middle n-gram as the first filtering pass to generate a list of candidates for incorrect words, if there is one candidate resulted from middle n-gram then no need for others filtering pass, and incorrect word will be replaced by this candidate. If else, the list will pass to the second filtering pass which is the *Levenshtein* distance algorithm, if there is one candidate has a unique smaller edit distance then no need to others filtering pass, and incorrect word will be replaced by this candidate. If else, the list will pass to the third filtering pass which is the left n-gram, and so on until the last n-gram, which is the right n-gram.

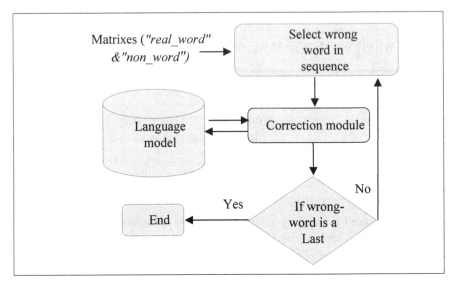

Fig. 5. Correcting of ASR text errors.

The candidates are generated by sending SQL query to corresponding n-gram table for each incorrect word and receive a table called *"candidates_list"* contains all these candidates with their frequencies. For example, let assume the incorrect-word is W[i], initially the program will send the sentence (W[$i-2$] W[$i-1$] % W[$i+1$] W[$i+2$]) to check their existence in corresponding n-gram table, where values of W[$i-2$], W[$i-1$], W[$i+1$] and W[$i+2$] must not exist in matrix *"real_word"* or matrix *"non_word"*, if exist then they will remove from sentence. Also, the symbol "%" represents any word that can come between W[$i-1$] and W[$i+1$]. Figure 6 shows another example of this process.

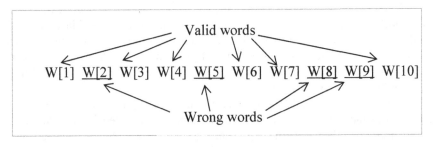

Fig. 6. Text contains ten words.

Figure 6 represents a text that contains ten words, where all the words are correct except underlined, it shows some cases of the position of errors in simple text. Now for the first incorrect word W[2], the sentence that needs to check became (W[1] % W[3] W[4]) and will check their existence in table "*gram4*". For the second incorrect word W[5] the sentence becomes (W[3] W[4] % W[6] W[7]) and will check their existence in table "*gram5*". For the third incorrect word W[8] the sentence becomes (W[6] W[7] %) and will check their existence in table "*gram3*". For the fourth incorrect word W[9]

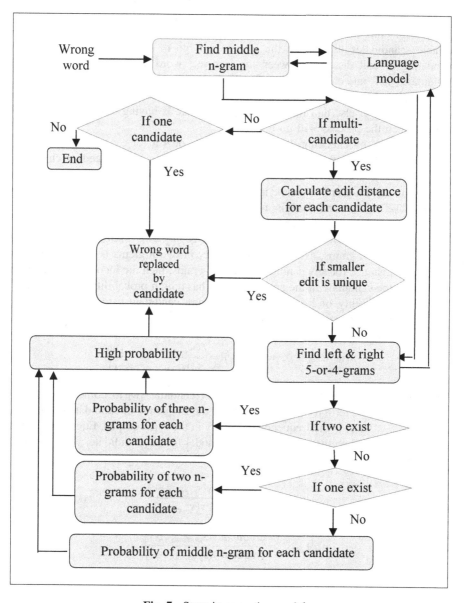

Fig. 7. Steps in correction module.

the sentence becomes (% W[10]) and will check their existence in table "*gram2*". After generating the list of candidates for each incorrect word, the process of correction is started. Figure 7 shows steps of correction of an incorrect word in the correction module.

Step 1: In case of no candidate exist in table "candidates_list", then the incorrect word will not be corrected in array "W". The correction will continue for other incorrect words. If else, then the process will go to step 2.

Step 2: In the case of one candidate exist in table "candidates_list". Then the incorrect word will be replaced by this candidate in the array "W". The correction will continue for other incorrect words. If else and there are more than one candidate exist in table "candidates_list". Then the process will go to step 3. Step 3 will start by calculating of edit distance between the incorrect word and all candidates in table "candidates_list" and check smaller edit distance if it is unique then the incorrect word will be replaced by this candidate in array "W". The correction will continue for other incorrect words. If else and more than one candidate is having the same smaller edit distance, then the process will go to step 4.

Step 4 starts by checking the existence of 5-g or 4-g for each candidate in table "candidates_list" on both sides that surround the incorrect word. It needs to check only 4–5-g because 3–2-g are checked in the middle n-gram. In the case of existing of one candidate in three n-grams, then the incorrect word will be replaced by this candidate in the array "W". If else and there is more than one candidate exists, Then the incorrect word will be replaced by the candidate in the array "W" that has a high probability in three n-grams. The priority for existence in three n-grams more than a priority in existence in two n-grams, even if the probability in two n-grams is better. If else and there is no existence for any candidate in two n-grams, then the incorrect word will be replaced by the candidate in the array "W" that has a high probability in the middle n-gram. The probability of language model for any candidate to come instead of an incorrect word can be calculated from Eq. 1:

$$p(w_i|w_{i-n}.\ldots.w_{i-2}w_{i-1}) = \frac{f(w_{i-n}.\ldots.w_{i-1}w_i)}{f(w_{i-n}.\ldots.w_{i-1})} \tag{1}$$

Where the symbol P is the probability for any candidate w_i to come instead of an incorrect word. The term $f(w_{i-n},\ldots,w_{i-1}, w_i)$ is the frequency of candidate w_i in corpus while $f(w_{i-n},\ldots,w_{i-1})$ is the frequency of history in the same corpus. Equation 1 can be represented the probability of three n-grams model for any candidate, which is called "c" to come instead of the incorrect word, as follow:

$$p(c)_{left} = \frac{frequency(s_{previous}\ c)}{frequency(s_{previous})} \tag{2}$$

$$p(c)_{middle} = \frac{frequency(s_{previous}\ c\ s_{next})}{frequency(s_{previous}\ \%\ s_{next})} \tag{3}$$

$$p(c)_{right} = \frac{frequency(c \; s_{next})}{frequency(s_{next})} \tag{4}$$

Where $P(c)_{left}$, $P(c)_{right}$ and $P(c)_{middle}$ are the probabilities of the left, right, and middle n-grams respectively. The symbol S represents a sentence consists of any 1-to-5-g. The terms $S_{previous}$ and S_{next} represent sentences coming before and after candidate "c" respectively. Hence, the total probability of three n-grams can be calculated from Eq. 5.

$$p(c)_{total} = \frac{p(c)_{left} + p(c)_{middle} + p(c)_{right}}{3} \tag{5}$$

The proposed language model in this research will not use any smoothing method because it will not apply through the process of converting speech to text but after the ending of this process. Hence, the proposed language model is used as an extra layer that is implemented after the text is generated. It will give the value of zero for any n-gram if their sentence is unseeing in any n-gram table.

4   Experimental Result

The experimental results of the proposed method have been evaluated against the results of two related methods. They are the Spectral Subtraction method [17] and Ensemble method [13, 18]. Furthermore, the experimental results have been performed using the *Kaldi* toolkit as an ASR system to convert speech to text. It is an open-source and free library used in several ASR domains [13, 19, 20]. In addition, this study used the same testing dataset that is used by [13]. Word Error Rate (WER) is used as a comparative metric in the evaluation process. The WER has been calculated using Eq. 6 [20, 21]. Table 2 shows the experimental results of the three tests using the metric Word Error Rate.

$$WER = \frac{Incorrect \; words}{Total \; words} \tag{6}$$

Table 2. Experimental results.

	Spectral subtraction	Ensemble method	The proposed method
Total words	4856	4856	4856
Incorrect words	2316	1663	900
WER	47.69%	34.24%	18.53%

The first row of Table 2 shows that the number of testing words in the output ASR text file is 4856 words. Row two shows the numbers of incorrect words that resulted from the Spectral Subtraction, Ensemble method, and the proposed method, which are 2316, 1663, and 900 respectively. The experimental results of the WER show different values for each tested method, Furthermore, the accuracy of all tested methods is still low for human speech recognition. In addition, the results also display that the Spectral Subtraction method achieved the highest value of WER with 47.69% followed by the Ensemble method with a WER of 34.24%. However, the proposed method was the one that achieved the best accuracy with a WER of 18.53%. It achieved a 15.71% relative reduction against the best WER for the comparative methods. Hence, the proposed method has been achieved the best accuracy against other related methods and it reduces the WER considerably.

5   Conclusion

This research presents a new method that can perform auto-detection and correction with considerable accuracy for incorrect words in ASR text. Furthermore, this method can be used for any language if its resources are available. It consists of two parts: the first part explains the steps to design and fill a language model database using a huge volume of data, and the other part shows the steps to implement the automatic correction of errors resulting from the ASR system. Data used to extract context information relied on Google n-gram files.

The experiment results show success in the extraction of context information from Google n-gram files. It also shows that using this method will reduce the word error rate considerably. Further research can be done for more improvement in the word error rate of ASR systems. Furthermore, it can try implementing this method in other languages and see the obtained results.

References

1. Paulraj, M., Yaacob, S., Yusof, S.M.: Vowel recognition based on frequency ranges determined by bandwidth approach. In: International Conference on Audio, Language and Image Processing, ICALIP 2008. IEEE (2008)
2. Marković, B., Galić, J., Grozdić, Đ., Jovičić, S.T., Mijić, M.: Whispered speech recognition based on gammatone filterbank cepstral coefficients. J. Commun. Technol. Electron. 62(11), 1255–1261 (2017). https://doi.org/10.1134/S1064226917110134
3. Bassil, Y., Alwani, M.: Post-Editing Error Correction Algorithm For Speech Recognition using Bing Spelling Suggestion. arXiv preprint arXiv:1203.5255 (2012)
4. Shahrul Azmi, M.Y., Nor Idayu, M., Roshidi, D., Yaakob, A.R., Yaacob, S.: Noise robustness of Spectrum Delta (SpD) features in Malay vowel recognition. In: Kim, T.-H., Ko, D.-S., Vasilakos, T., Stoica, A., Abawajy, J. (eds.) FGCN 2012. CCIS, vol. 350, pp. 270–277. Springer, Heidelberg (2012). https://doi.org/10.1007/978-3-642-35594-3_38
5. Abdulkhudhur, H.N., et al.: Implementation of improved Levenshtein algorithm for spelling correction word candidate list generation. J. Theor. Appl. Inf. Technol. 88(3), 449–455 (2016)

6. Jurafsky, D., et al.: Speech and Language Processing: An Introduction to Natural Language processing, Computational Linguistics, and Speech Recognition, vol. 2. MIT Press, Cambridge (2000)
7. Abdulkhudhur, H.N., et al.: The UX of banking application on mobile phone for novice users. J. Comput. Theor. Nanosci. **16**(5–6), 2218–2222 (2019)
8. Kaki, S., Sumita, E., Iida, H.: A method for correcting errors in speech recognition using the statistical features of character co-occurrence. In: Proceedings of the 36th Annual Meeting of the Association for Computational Linguistics and 17th International Conference on Computational Linguistics-Volume 1. Association for Computational Linguistics (1998)
9. Mangu, L., Padmanabhan, M.: Error corrective mechanisms for speech recognition. In: 2001 IEEE International Conference on Acoustics, Speech, and Signal Processing. Proceedings (ICASSP 2001). IEEE (2001)
10. Jeong, M., Jung, S., Lee, G.G.: Speech recognition error correction using maximum entropy language model. In: Proceedings of INTERSPEECH (2004)
11. Zhou, Z., Meng, H.M., Lo, W.K.: A multi-pass error detection and correction framework for Mandarin LVCSR. In: INTERSPEECH (2006)
12. Ringger, E.K., Allen, J.F.: Error correction via a post-processor for continuous speech recognition. In: 1996 IEEE International Conference on Acoustics, Speech, and Signal Processing, ICASSP-96. Conference Proceedings. IEEE (1996)
13. Habeeb, I.Q., et al.: An ensemble technique for speech recognition in noisy environments. Indones. J. Electr. Eng. Comput. Sci. **18**(2), 835–842 (2020)
14. Habeeb, I.Q., et al.: Constructing Arabic language resources from Google N-gram dataset. J. Phys.: Conf. Ser. (2020)
15. Attia, M., et al.: An automatically built Named Entity lexicon for Arabic. In: Proceedings of the 7th conference on International Language Resources and Evaluation (LREC 2010), Valletta, Malta (2010)
16. Remy, M.: Wikipedia: the free encyclopedia. Ref. Rev. **16**(6), 5 (2002)
17. Puligilla, S., Mondal, P.: Co-existence of aluminosilicate and calcium silicate gel characterized through selective dissolution and FTIR spectral subtraction. Cem. Concr. Res. **70**, 39–49 (2015)
18. Habeeb, I.Q., Al-Zaydi, Z.Q., Abdulkhudhur, H.N.: Enhanced ensemble technique for optical character recognition. In: Al-mamory, S.O., Alwan, J.K., Hussein, A.D. (eds.) NTICT 2018. CCIS, vol. 938, pp. 213–225. Springer, Cham (2018). https://doi.org/10.1007/978-3-030-01653-1_13
19. Povey, D., et al.: The Kaldi speech recognition toolkit. IEEE Signal Processing Society (2011)
20. Wang, Z., et al.: Rank-1 constrained multichannel Wiener filter for speech recognition in noisy environments. Comput. Speech Lang. **49**, 37–51 (2018)
21. Habeeb, I., Al-Zaydi, Z., Abdulkhudhur, H.: Selection technique for multiple outputs of optical character recognition. Eurasian J. Math. Comput. Appl. **8**(2), 41–51 (2020)

Networks

A Proposed Dynamic Hybrid-Based Load Balancing Algorithm to Improve Resources Utilization in SDN Environment

Haeeder Munther Noman[1(✉)] and Mahdi Nsaif Jasim[2]

[1] Software Department, College of Information Technology,
University of Babylon, Babylon, Iraq
[2] University of Information Technology and Communications, Baghdad, Iraq

Abstract. Several load balancing schemes are proposed to tackle the webserver overloading problems. Static load balancing is appropriate for systems with low load variations where the traffic fairly distributes among servers and Prior information about system resources is required. Dynamic load balancing monitors the system's current state to perform load controlling actions and respond to the current system state while making load transferring decisions. Consequently, processes may dynamically switch from an overused machine to underuse during real-time. However, to utilize the resources more efficiently this paper proposes a new hybrid-based load balancing algorithm that relies on inheriting the distinctive characteristics, overcoming the existing limitations, and combining the desired features of static as well as dynamic load balancing. The experimental analysis witnessed the utilization of the OpenLoad benchmarking tool that generate a concurrent users from 0 up to 350 to provide a near real-time performance measurement of the application under test and to evaluate the performance of load balancing algorithms. Results reveal that proposed hybrid-based load balancing algorithm interestingly enhances server transactions per second up to 10.41%, average server response time up to 24.61%, and server CPU capacity up to 9.55% when compared with other load balancing algorithms like static weighted round-robin (WRR) and dynamic least connection-based (LCB). Accordingly, this research recommends deploying the proposed load balancing technique in SDN-Based Platform data center networks (DCN's).

Keywords: Weighted Round-Robin (WRR) · Least Connection-Based (LCB) · Software Defined Network (SDN) · Advanced Server Monitoring module (ASM) · Hybrid-based (HB)

1 Introduction

SDN is a modern architecture that separates control and forwarding functions in the network. The separation is a departure from the conventional architecture where complex tasks are abstracted from the repeated forwarding tasks. A complex process is automated and handled separately in a centralized SDN controller or Network Operating System. The controller uses the OpenFlow protocol to connect with the real physical or virtual switch, and the data path uses the flow entries inserted by the

© Springer Nature Switzerland AG 2021
A. M. Al-Bakry et al. (Eds.): NTICT 2021, CCIS 1511, pp. 147–162, 2021.
https://doi.org/10.1007/978-3-030-93417-0_10

controller in the flow table for routing data. Three methods exist to separate the control plane and the data plane: fully distributed, logically centralized, and strictly centralized. In the Fully distributed method, switching devices provide one essential feature for forwarding packets, but unfortunately, without a control power, that may lead to a failure point. The logically centralized method includes a remarkable feature signified in the devices with a partial functionality embedded inside. The strictly centralized method adopts the conventional way of making all machines within all planes route packets through the network. However, the majority of clients, organizations, and businesses rely on the internet for their daily activities to deliver services online [1]. The hardware performance improvements are no longer enough to cope with the rising volume of customer requests while preserving the desired quality of service [2]. Accordingly, standard practice is to utilize servers to process the client requests [3]. Suppose it is difficult to allocate the incoming requests among servers evenly. In that case, some servers could be overloaded while others remain idle, which leads to low server utilization and poor service quality. Moreover, due to the ever-increasing request loads of popular websites, it's challenging to adopt a single robust server or mirrored servers [4]. Therefore, a load balancer (also known as a dispatcher) is employed to distribute client requests to a particular server in the back end and removes any possible single point of failure to maximize system reliability [5]. The load balancing strategies invoked by the load balancer are viable to redirect requests among the server members and become necessary when multiple servers operate simultaneously. Moreover, load balancing techniques reduce the overall response time and optimize the total throughput when all activities are transparent to the user [6]. The contribution is to develop and implement a hybrid-based load balancing algorithm that eliminates the issues and inherits the merits of static and dynamic load balancing schemes.

The organization of this paper is as follows: Sect. 2 introduces a literature review related to the development of load balancing algorithms running in SDN, the main features of static and dynamic load balancing algorithms, and the limitations for both types of algorithms. Section 3 outlines the proposed hybrid-based (HB) load balancing algorithm. Section 4 discusses the experimental set up. Section 5 is committed to discuss the flow-sequence diagram of the proposed algorithm. Section 6 describes the experimental results, evaluation, and finally Sects. 7 and 8 deal with the conclusion and future work.

2   Related Work

Many approaches have been hypothesized to address the load balancing issue. Traditional load balancing algorithms suffered from being non-programmable and vendor locked, which turned them to be rendered inoperable. Moreover, network administrators could not build customized applications [7]. Preliminary work focused on a single load balancing parameter was later insufficient to pick up the best server for processing user requests [8]. A systematic study to investigate a dynamic flow entry saving multipath (DFSM) system for inter-data center WAN transmission was carried out [9]. The DFSM system employed the concept of source-destination multipath forwarding mechanism with latency awareness flow-based traffic splitting to preserve flow entries and get a high level of achieved performance. However, the DFSM saved from 15% up to 30% of system flow entries and reduced from 10% up to 48% of average latency. The computation of the shortest route among hosts and performing the load calculations associated with each link was proposed by [8]. If congestion in a route occurs, it substitutes the old route with the alternative best route having the lowest traffic flow [10]. The utilization of the dispatcher architecture represented the major advance regarding load balancing to employ the back-end server, later investigated to be inappropriate for distributing client requests evenly to different servers [11]. The hybrid–based load balancing algorithm's proposition relying on each of the Least Load and Round-Robin load balancing solutions running in an SDN environment was suggested by [12]. Accordingly, the time has come to develop different hybrid load balancing algorithm that merges two or more load balancing schemes, static or dynamic. The aim is to produce a new algorithm that incorporates the benefits, inherits the distinctive characteristics, and overcomes both strategies' limitations and inconveniences.

2.1 Static Load Balancing Algorithms

An equal division of traffic is performed inside the servers in a static algorithm, which is appropriate for platforms where the load changes at a low rate. Prior information about system resources is required to ensure that load shifting decisions do not rely on the system's current state. Moreover, initial tasks are assigned to the individual processors for execution by the master processor [13]. Accordingly, the performance of the workload is calculated from the beginning through the master processor. Slave processors calculate the expected work and provide the results for the master processor. Static load balancing is appropriate for systems with low load variations where the traffic fairly distributes among servers. The processor's efficiency is computed at the start of execution, and the decision to transfer loads avoids relying on the system processors' current status. Tasks are assigned to virtual machines and processors after their generation because they could not be transferred to any other device for load balancing during the execution. Although a static scheme is less overhead, suitable for a homogeneous environment, and easy to implement, several drawbacks are reported: like the non-versatile nature, the inability to accept dynamic changes, pure dependency on statically obtained data, and the non-pre-emptive behavior. WRR is a good instance of this type of algorithm regarded as the Round-Robin's improved version. Static weight is assigned to each server in the network pool based on servers' actual capacities and specifications regarding CPU, RAM, etc. [14].

2.2 Dynamic Load Balancing Algorithms

Due to the dynamic distribution of the pre-programmed load balancing patterns, the dynamic load balancing scheme is more efficient than static [15]. Two modes are available to implement the dynamic load balancing scheme: the non-distributed and distributed. In the nondistributed approach, a (centralized) node receives and distributes all requests to the servers, whereas the distributed mode shares the nodes within the distribution of the requests [16]. Features of dynamic load balancing schemes include monitoring the system's current state to perform load controlling actions and responding to the current system state while making load transferring decisions. Consequently, processes may dynamically switch from an overused machine to an underused device in real-time. Dynamic load balancing schemes come with several advantages, like the absence of a single web server overloading problem. The current system load is kept under consideration to select the following data center. Drawbacks include communication overheads, the growing number of procedures, and the higher run time complexity. However, LCB is one of the traditional dynamic load balancing schemes in circulation. The load distribution is decided based on the current number of connections for every node. A load balancer keeps a log of a node's number of connections. However, the node with the fewest number of connections is firstly selected [17] such that when a new operation or order occurs, load increases, whereas when the operation terminates, load decreases. This technique is a good choice if the request load has a high degree of variance and is ideal for an individual server pool with a similar capacity for each member node.

2.3   Limitations and Drawbacks

Load balancing schemes perform well but unfortunately hold certain limitations. WRR is simple and runs fast but lacks accuracy during load balancing of varying load size or request complexity. Current server load is not considered during request distribution [18]. In the LCB algorithm, the server load is taken into account before load delivery, which may have the load distributed sparsely, leaving some of the servers idle [19].

3   Proposed Hybrid-Based (HB) Algorithm

The proposed load balancing algorithm consists of two functional models, the Advanced Server Monitoring Module (ASMM) and the Hybrid Load Balancing Module (HLBM). The modules usually mount on the top of the POX controller. The ASMM module tracks all of the information related to the server's status, modify some of the corresponding parameters, and forwards it periodically to the load balancing module. The Hybrid load balancing Module (HLBM) is considered the main operating module responsible for handling and controlling the load balancing decisions. Each server connects to the OpenFlow switch with a static IP address, and each server pool acquires a virtual IP and MAC address. Without recognizing the server's physical address, users send their requests to the OpenFlow switch's virtual MAC address, which sends a Packet_in message to the controller running the modules mentioned above. If there is no matching regarding flow entries, the controller inserts the corresponding flow entry into the OpenFlow switch via the southbound OpenFlow protocol.

3.1   Advanced Server Monitoring Module (ASMM)

The ASMM is developed to collect the OpenFlow switch ports, flow entries, and flow table counters statistics. The OpenFlow switch provides three types of counters statistics: the first type is the statistics per flow table that consist of matched packets, the number of looked-up packets, and the number of active flow entries. The second type is the statistics per flow entry in a flow table that consists of the received bytes, received packets, and duration. The last type is the statistics per port consisting of received packets, transmitted packets, received bytes, transmitted bytes, receive drops, transmit drops, receive errors, transmit errors, and collisions. The module executes a runnable class, which involve the running data collection function according to the time interval set to 5 s by passing out two parameters: the first parameter is the OpenFlow switch port number (OpenFlowPortNo.) used to retrieve the counters statistics related to OpenFlow switch ports such as the received packets, transmitted packets, received bytes, transmitted bytes, received drops, transmitted drops, and the collisions. The second parameter is the Flow entry number (FlowEntryNo.) is used to retrieve the counters statistics related to flow entries such as received bytes, received counters, time and OpenFlow table statistics such as active entries, packet lookups, and packet matches. The ASMM is incorporated into the HLB by using a thread that handles the request and response statistics as explained in Algorithm 1.

Algorithm1: Advanced Server Monitoring Module(ASMM)

Input: *Network topology, O.FSwitch PortNo, O.FSwitch FlowEntryNo, t //adopted to collect OpenFlow Switch traffic information requested by POX*
Output: *Collect OpenFlow switch counters statistics periodically every 5 seconds*

SwitchStatistics: SwitchStatisticsType
While *TRUE do*
    Read *SwitchStatisticsM*
    Count SwitchStatisticsM.Port for the Server Connected to the OpenFlow Switch by OfPortNo.
    Count SwitchStatisticsM.FlowEntry for the Server Connected to the OpenFlow Switch by OfPortNo
    Update *SwitchStatisticsM.FlowTable, SwitchStatisticsM.FlowEntry, SwitchStatisticsM.Port*

    Sleep (t units of time)

End while

3.2 Hybrid Load Balancing Module (HLBM)

This section aims to explain the approach followed by this module to compute and distribute the incoming traffic to the server. The load is determined based on least connections as well as static weight for each server. The requests usually demand pages of type text with small sizes as the HTTP web server itself could process them and do not require many resources. Thus, consuming less bandwidth and avoiding server-side scripting (e.g., JSP, ASP, and PHP). In contrast, requests demanding pages of big sizes like images that require additional resources cause extra time and consume high bandwidth. The HLBM implements the load of each server and selects the optimal server to handle the incoming request, as shown in the processes listed below:

1. **Computing the Load of Servers**

   The module adopts a single pool (P) assigned with a single service type (HTTP). The pool includes server members acquiring varying load, L = (L(S 1), L (S2)....L (SN)) and R is the set of requests that need to be scheduled: R = (R1, R2....RN). The load of the pool is calculated according to Eq. 1:

$$L(Si)Pool = \sum_{i=0}^{n} L(Si) \tag{1}$$

   The load of the server members is calculated based on the algorithmic rule that consists of two stages. The first stage continuously monitors server members and computes the active number of connections of each server, and selects the server with the least number of real-time active connections according to Eq. 2:

$$Sm = Min[Conn \sum_{i=1}^{n} Si] \tag{2}$$

   Where Sm: refers to the server member having the least number of connections. However, the second stage carries out the mathematical multiplication of least connections server member resulting from the first condition with static weight for each server in the pool. The server producing the highest product value, is selected and the request is forwarded to that server depending on Eq. 3:

$$Soptimal = Max[\sum_{i=1}^{n} (Sm) * W(Si)] \tag{3}$$

   Where Soptimal: refers to the final server member responsible for handling the incoming request among a pool of n servers. n, i = {1, 2, 3,}, where n is the total number of servers in the pool. W (Sw): refer to the assigned static weight for each server member.

2. **Selecting the Optimal Server**

   As the POX controller receives requests from clients, then relying on the calculations of the mathematical rule mentioned in the above step, the server with the least load is selected to process the client request. However, each time a load of server members varies, then related information is updated as explained in Algorithm 2.

Algorithm2: Hybrid–Load Balancing Module (HLBM)

Input: *ServersIDArray = {S1, S2… Sn}, ServersWeightArray = {ServerWeight (1), ServerWeight (2),…..ServerWeight(n)*

Output: *The Best server to Handle the Incoming Request Among a Pool of n servers*

Type1: Type

    ServerNo.: integer

    ServerNo.of.Connections: integer

    ServerWeight: integer

    OptimalServer: integer

    SelectServNo: integer

    No.ofServers: integer

end Type1

Servers: Array[No.of Servers] of Type1

OptimalServers: Array[No.ofServers]

Type2: *SwitchStatisticsType*

Record Per FlowTable

    ActiveEntries: Integer

    PacketLookups: Integer

    PacketMatches: Integer

End Record

Record PerFlowEntry

    ReceivedPackets: Integer

    ReceivedBytes: Integer

End Record

Record PerPort

    ReceivedBytes: Integer

    TransmitBytes: Integer

    ReceivedPackets: Integer

    TransmitPackets: Integer

    ReceiveDrops: Integer

    TransmitDrops: Integer

    Collisions: Integer

End Record

end Type2

SwitchStatistics: SwitchStatisticsType

Call Advanced Server MonitoringModule(SwitchStatistics)

for *i=1 to No.of Servers*

    Call FindMinServerConn. (No. of Servers, Servers, and Var SelectedMinServerConn.

    MinConnServer = SelectedMinServerConn

End for

for *i=1 to No.of Servers*

    OptimalServer[i] =Server[i].weight* MinConnServer

end for

SelectedServNo=findMax(OptimalServers)

forward the request to (SelectedServNo)

Algorithm FindMinServerConn.(N, Servers, Var SelectedMinServConn.)

N= Number of Servers

ServerMinConnections:int=0

for *m = 0 to n-1*

    If *Servers[m].Conn. <ServerMinConn.*

      Then *ServerMinConn. = Servers[m].conn.*

    end if

end for

Return *ServerMinConn*

End.

4 Experimental Setup

The experimental work is carried out using a POX controller. The POX is written with Python, designed and implemented as a friendlier alternative by several SDN developers and Engineers. POX performs well as compared to NOX. The POX is a more straightforward development environment to deal with that offers a web-based GUI with a fairly well-written API and documentation. Moreover, the POX facilitates the addition and removal of restored units, and enhances the experiments flexibility and performance. The POX includes a list of IP addresses assigned to each server statically. The Mininet emulator carries creates the virtual network topology that consists of three hosts acting as Apache HTTP web servers holding the same configuration to provide the same web services to the clients. Nine hosts acting as clients, which attempt to access the Apache HTTP web servers by using a virtual IP address. The OpenLoad benchmarking tool plays an important role in evaluating the performance of the proposed load balancing scheme. OpenLoad is easy to use and provides a near real-time performance measurements of the application under test. This is particularly useful during performing optimization due to the fact that OpenLoad reflects the impact of modifications almost immediately. This tool examines the impact of the gradual growth of load represented by concurrent users from 0 up to 350 on the server's quality of service (QoS) metrics defined by the server's average response time, server's transactions per second, and server's CPU capacity to handle requests. Port 6633 is the default connection port for establishing communications between the OpenFlow switch and POX controller.

5 Flow Sequence Diagram

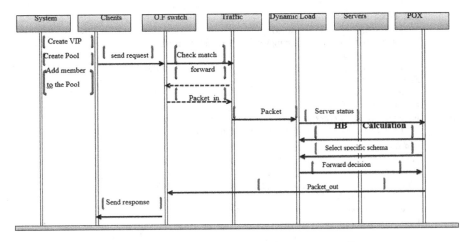

Fig. 1. HB load-balancing algorithm flow-sequence diagram

As shown in Fig. 1, the flow-sequence diagram Steps of HB load balancing algorithm is listed as follows:

- **System Configuration Process**
  The user should initially set up the server pool, the VIP address associated with that pool, and the traffic type. The configuration of the server pool is based on the number of member servers. Since a single service is provided by the server members and single type of transmission that is TCP. Therefore, the proposed system requires a single pool
- **The Addition of Member Servers to Pool**
  Three HTTP servers (HTTP Server1, HTTP Server2, and HTTP Server3) are launched into the pool relying on their layer three static IP address with the same port of VIP.
- **Sending Requests**
  The OpenLoad benchmarking tool plays an essential role in evaluating the performance of server members. OpenLoad is discovered to be flexible for use and provides a near real-time performance measurement of the application under test, which is particularly useful when performing optimization as the impact of changes could be immediately observed. This tool relies on investigating the effect of the gradual growth of load represented by concurrent users on the server's performance metrics. The number of concurrent users' requests varies from 0 up to 350 (req/sec) with a uniform increase of 50 (req/sec)
- **Receive and Process Packet_in Messages**
  When the message reaches the OpenFlow switch and no base entry is found, then it is forwarded to the POX controller, which verifies whether the message is IPv4 or not. If yes, then the packet is parsed to obtain details like the type of service, destination IP, and source IP. The POX controller utilizes the IP address of server members and the VIP to send ARP messages in order to check if there is a server crash (the server that has not responded to this message for some time) and must be deleted from the server list in the load balancing application.
- **Load Balancing Mechanism**
  The ASMM reports the OpenFlow switch statistics, including ports and flows entry counter statistics of each server member in the pool periodically every 5 s. The HLBM executes the tasks represented by the computation of the load for each server member according to Eq. (4) and Eq. (5) to select the optimal server for handling the incoming request from the client. The experimental evaluation compares the proposed HB load balancing algorithm with two dynamic-based load balancing algorithms represented by WRR and LCB under the same conditions. WRR is simple and runs fast but lacks the accuracy during load balancing of varying load size or request complexity, and current server load is not considered during request distribution. The LCB takes into account the server load before load delivery, which ends up when the load is acquired sparsely, leaving some servers idle beside the Addressing the imperfect system performance when the processing capabilities of the servers are different, thereby reducing system performance (Table 1).

Table 1. Experimental parameters values.

Parameter	Value
Host operating system	Linux (Ubuntu) version 14.04
Programming language	Python 2.7.6
Simulation tool	OpenLoad
Max no. concurrent users	350
Min no. concurrent users	0
No. of servers (3)	3
No. of clients	9
SDN controller	(POX)
Virtual SDN switch	OpenFlow switch
POX Controller to OpenFlow switch port	6633
HTTP web servers listening port	80
Virtual IP (service IP)	10.0.1.1
No. of iterations	7
Link Latency	No

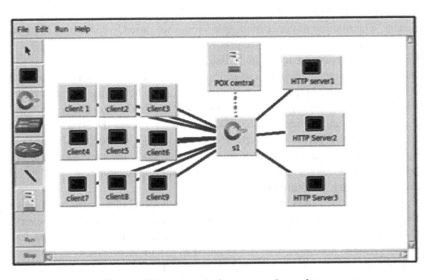

Fig. 2. SDN-based platform network topology

The network topology in Fig. 2 utilizes a single remote POX due to owning free, open-source utilities that facilitate the addition and removal of restored units and allow the experiments to be more flexible and reliable. Moreover, this section adopts an OVS switch along with a companion Linux kernel module for flow-based switching as shown in Fig. 2. The listening port 6633 is the default connection port for the mininet to create connections between the OpenFlow switch and the POX. The Mininet emulator creates a virtual network topology that consists of 3 hosts acting as

Apache HTTP web servers that offer the same web services and 9 hosts' serves as clients who access the web servers using a virtual IP address.

6 Results and Discussions

The outcomes of the research can be discussed as follows:

6.1 Server Transactions per Second

The Server Transactions per Second is the first metric to investigate, which is regarded as one of the essential performance parameters capable of handling routine and keeping records. In general, this metric defines the total number of transactions accomplished by a server in a given period divided by the seconds per that period. The load is represented by the simultaneous number of users attempting to access the local server webpage, which starts from 0 up to 350 to verify the ability of load balancing schemes to distribute several flows in a parallel path between the source and the destination.

$$\text{Transactions per Second (TPS)} = \frac{TPS1 + TPS2 \cdots \ldots TPSn}{\sum T} \tag{4}$$

Related to Eq. (4), Fig. 3 Demonstrates servers' transactions per second when adopting WRR, LCB, and proposed HB load balancing schemes. The experimental results indicate a noticeable improvement to the proposed HB scheme, recording 67.73 compared to 53.5, 60 to WRR and LCB schemes, respectively.

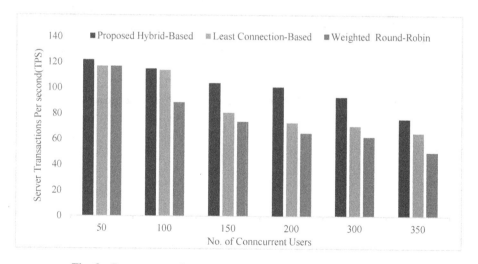

Fig. 3. Server transactions per second vs. the no. of concurrent users

6.2 Server Average Response

This parameter expresses the required time to deliver request results to the clients. Different variables may affect the response time, such as average thinking time, number of requests, number of users who accessed the system, and the bandwidth [20].

$$\text{Server Average Response Time (SART)} = \frac{Un}{Rn} - Tt. \tag{5}$$

Whereas T_T is the thinking time per request, Un is the number of concurrent users, and Rn is the number of requests per second. According to Eq. (5), Fig. 4 proves that polylines fluctuate in a stable and minimal form during the application of the proposed HB scheme, which leads to selecting the server with the minimum average response time of 1.28 ms as compared with 1.49, 1.51 ms associated to WRR and LCB load balancing policies (Fig. 5).

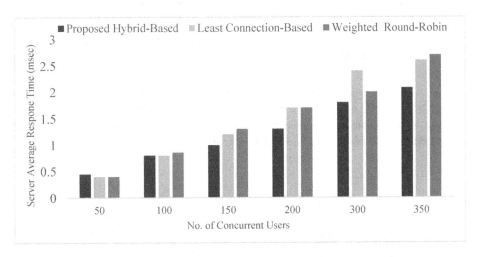

Fig. 4. Server Average Response Time vs. the no. of concurrent users

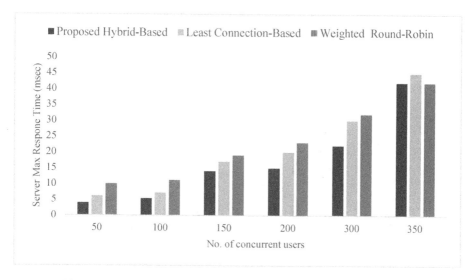

Fig. 5. Server Maximum Response Time vs. the no. of concurrent users

6.3   Server CPU Capacity

The Server CPU capacity is the final metric to investigate, which refers to the process responsible for calculating the number of resources needed to provide the desired level of services for a given workload, as presented in Eq. (6) [20].

$$\text{Server CPU capacity} = \frac{1}{SART} \tag{6}$$

According to Eq. (6), the results from Fig. 6 unmistakably imply that the proposed HB scheme records the highest ratio of Server CPU capacity up to 1.033 compared to 0.988 and 0.981 belonging to the LCB and WRR load balancing schemes.

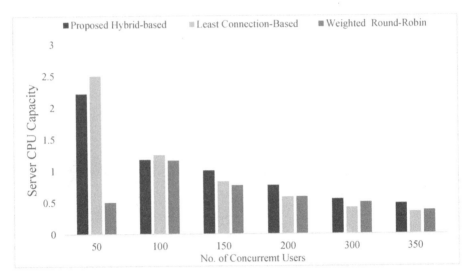

Fig. 6. Server CPU Capacity vs. the no. of concurrent users

7 Conclusion

The development of SDN architecture offered new possible options to solve the conventional load balancing problems. This article suggested a new load balancing scheme using the POX controller under the SDN architecture to pick the server, achieving several vital metrics in SDN like the maximum transactions per second, minimum average response time, and maximum server CPU capacity. Moreover, it seemed to be sufficient to overcome low-performance problems and high load balancing costs. It is possible to conclude that the proposed HB load balancing algorithm achieves a significant progress in terms of server transactions per second up to 10.41%, server average response time up to 24.61%, and server CPU capacity up to 9.55%. It is also essential to state that any attempt to increase the number of concurrent users above 350 lead servers to enter the saturation region, which causes several drawbacks and performance degradation.

8 Future Work

In the following, we present possible future research directions that may be conducted to extend the dissertation innovations:

- Future work will address the scalability of the proposed HB load balancing algorithm achieved by the dynamic addition of the hosts to the existing pool when all pool members are being overloaded.

- Adopting multiple controllers to avoid the single point of failure problem as additional controllers carry out the load balancing task when the master controller goes down.
- The assignment of dynamic weights instead of static ones for the servers participating in the proposed hybrid-based load balancing scheme. Dynamic weights generally reflect actual server capabilities and further improve the performance of the balancing scheme.

References

1. Trestian, R., Katrinis, K., Muntean, G.M.: OFLoad: an OpenFlow-based dynamic load balancing strategy for datacenter networks. IEEE Trans. Netw. Serv. Manag. **14**(4), 792–803 (2017)
2. Semong, T., Maupong, T., Anokye, S., Kehulakae, K., Dimakatso, S., Boipelo, G., Sarefo, S.: Intelligent load balancing techniques in software defined networks: a survey. Electronics **9**, 1091 (2020)
3. Xie, J., Guo, D., Hu, Z., Qu, T., Lv, P.: Control plane of software defined networks: a survey. Comput. Commun. **67**, 1–10 (2015)
4. Mendiola, A., Astorga, J., Jacob, E., Higuero, M.: A survey on the contributions of software-defined networking to traffic engineering. IEEE Commun. Surv. Tutor. **19**(2), 918–953 (2016)
5. Kavana, H.M., Kavya, V.B., Madhura, B., Kamat, N.: Load balancing using SDN methodology. Int. J. Eng. Res. Technol. **7**, 206–208 (2018)
6. Bholebawa, I.Z., Jha, R.K., Dalal, U.D.: Performance analysis of proposed OpenFlow-based network architecture using Mininet. Wirel. Pers. Commun. **86**, 943–958 (2016)
7. Kreutz, F.M., Ramos, V., Verissimo, P.E., Rothenberg, C.E., Azodolmolky, S., Uhlig, S.: Software-defined networking: a comprehensive survey. Proc. IEEE **103**, 14–76 (2014)
8. Chen-Xiao, C., Ya-Bin, X.: Research on load balance method in SDN. Int. J. Grid Distrib. Comput. **9**, 25–36 (2016)
9. Muthumanikandan, V., Valliyammai, C.: Link failure recovery using shortest path fast rerouting technique in SDN. Wirel. Pers. Commun. **97**, 2475–2495 (2017)
10. Chen, L., Qiu, M., Dai, W., Jiang, N.: Supporting high-quality video streaming with SDN-based CDNs. J. Supercomput. **73**(8), 3547–3561 (2016). https://doi.org/10.1007/s11227-016-1649-3
11. Karakus, M., Durresi, A.: A survey: control plane scalability issues and approaches in software-defined networking (SDN). Comput. Netw. **112**, 279–293 (2017)
12. Jha, R.K., Llah, B.N.: Software Defined Optical Networks (SDON): proposed architecture and comparative analysis
13. Sufiev, H., Haddad, Y., Barenboim, L., Soler, J.: Dynamic SDN controller load balancing. Future Internet **11**, 75 (2018)
14. Mehra, M., Maurya, S., Tiwari, N.K.: Load balancing in software defined network: a survey. Int. J. Appl. Eng. Res. **14**, 245–253 (2019)
15. Chahlaoui, F., Dahmouni, H.: A Taxonomy of load balancing mechanisms in centralized and distributed SDN architectures. SN Comput. Sci. **1**, 1–16 (2020)
16. Iyer, N., Hugar, N.S., Manjunath, M.N.: Load balancing using open daylight SDN controller: case study. Int. Res. J. Adv. Sci. Hub **2**, 59–64 (2020)

17. Jader, O.H., Zeebaree, S.R., Zebari, R.R.: A state of art survey for web server performance measurement and load balancing mechanisms. Int. J. Sci. Technol. Res. (IJSTR) **8**, 535–543 (2019)
18. De Rango, F., Inzillo, V., Quintana, A.A.: exploiting frame aggregation and weighted round robin with beamforming smart antennas for directional MAC in MANET environments. Ad Hoc Netw. **89**, 186–203 (2019)
19. Singh, N., Dhindsa, D.K.: Hybrid scheduling algorithm for efficient load balancing in cloud computing. Int. J. Adv. Netw. Appl. **8** (2017). 0975-0290
20. Osman, A.A.A.: Service based load balance mechanism using Software-Defined Networks. Doctoral dissertation, University of Malaya (2017)

Energy-Saving Adaptive Sampling Mechanism for Patient Health Monitoring Based IoT Networks

Duaa Abd Alhussein[1], Ali Kadhum Idrees[1]([envelope]) [iD],
and Hassan Harb[2] [iD]

[1] Department of Computer Science, University of Babylon, Babylon, Iraq
ali.idrees@uobabylon.edu.iq
[2] Computer Science Department, American University of Culture
and Education (AUCE), Nabatiyeh/Tyre, Lebanon
hassanharb@auce.edu.lb

Abstract. This paper proposes an Energy-saving Adaptive Sampling Mechanism (EASaM) for patient health monitoring in the Internet of Things (IoT) networks. EASaM is implemented in the biosensors to remove redundant data during monitoring the status of the patients. It operates in the way of rounds. There are two periods in the round. The emergency discovery and adapting the sampling rate of each biosensor are two main steps in EASaM. The NEWS is implemented at each biosensor to eliminate the repetitive sensed data before forwarding it to the coordinator. The sampling rate is modified after every two periods based on the status of the patient at the end of each round. We achieved several experiments based on real sensed data from the biosensors of the patients. The results explain that the proposed EASaM decreased the sent data up to 80.75% and 85% for the high-risk patient and low-risk patients and the energy consumption is decreased whilst keeping a good representation for the whole scores at the coordinator in comparison with other methods.

Keywords: Wireless Body Sensor Networks (WBSNs) · IoT · Adaptive sampling · Patient health monitoring · Emergency detection

1 Introduction

In recent years, the world faces an increasing number in the illness and patients. Moreover, wars and the relationships between the human and animals led to introduce and spread new kinds of viruses and unknown diseases such as coved-19. Consequently, this will make the work of doctors and nurses is very difficult [1].

Last years, providing health care for patients has received many advantages from governments and the world. They spent high costs to provide different services of health and applications [2]. The rapid advancement of IoT technology, medical sensors, and huge data strategies led to increase and emerge the healthcare systems that called connected healthcare [3, 4]. This technology allowed professionals to access patient data remotely and provided simple and inexpensive options for monitoring and tracking the patients wherever they were and at any time.

© Springer Nature Switzerland AG 2021
A. M. Al-Bakry et al. (Eds.): NTICT 2021, CCIS 1511, pp. 163–175, 2021.
https://doi.org/10.1007/978-3-030-93417-0_11

Typically, the healthcare systems are composed of biosensors that provide patients' monitoring. These biosensors collect the vital signs (of oxygen rate, respiration rate, oxygen rate, blood pressure, heart rate, and temperature) for the patients continuously and periodically and then send these sensed vital signs to the gateway for larger processing [3, 5]. The connected healthcare applications are facing some important challenges like saving the power of the biosensor devices to ensure a long time monitoring as possible for the patients, and speed up discovering of the patient's emergency and send it to the medical specialist to provide the suitable decision.

To deal with these challenges, this paper suggests an Energy-saving Adaptive Sampling Mechanism (EASaM) for patient health monitoring in the Internet of Things (IoT) networks. EASaM is executed in the medical biosensors to eliminate unnecessary data during controlling and observing the situation of the patients. It functions in rounds. The round includes two periods. The emergency detection and modifying the sampling rate of each biosensor are two main steps in EASaM. The NEWS is achieved at each biosensor to drop the repeated data before delivering it to the gateway. The sampling rate is adjusted after every two periods based on the situation of the patient at the end of each round.

The rest of this paper is arranged as follows. The next section explained the related work. Section 3 demonstrates the proposed EASaM Approach for Smart Healthcare in IoT networks. The simulation results and analysis are presented in Sect. 4. Section 5 introduces the conclusions and future work. The healthcare system is composed of a group of biosensors that are responsible for monitoring the patient situation and sensing the vital signs like respiration, oxygen, rate of heart, the pressure of blood, temperature, etc., then send it to the coordinator to achieve more analysis and processing [3, 5].

There are some important challenges in the Connected healthcare application such as: decrease the consumed energy of the biosensor devices to guarantee as long monitoring as possible for the patient and fast detecting of the patient's emergency and report it to the medical experts to provide the appropriate decision.

To deal with these challenges, this paper proposed an Energy-efficient Adaptive Sensing technique (EASeT) for Smart Healthcare in IoT networks. EASeT integrate between two energy saving approaches: data reduction with emergency detection of patient and adaptive sensing of the biosensor. The first phase aims to discover the emergency of the patient and eliminate the repetitive medical data before send it to the coordinator. The second phase achieves the adaptive sensing based on the similarity between the scores of the last two periods. The remainder of this paper is organized as follow. The related work is explained in the next section. Section 3 demonstrates the proposed EASeT technique for Smart Healthcare in IoT networks. The simulation results and analysis are presented in Sect. 4. Section 5 introduces the conclusions and future work.

2 Related Literature

One of the effective solutions in hospitals is to use the connected healthcare to save and process the sensed vital signs of the patients to make the appropriate decision to save their lives. Some related work is focused on compression methods to reduce the huge data [6, 7], aggregation [8, 9] and prediction methods [10]. In [10], the authors propose a technique named PCDA (Priority-based Compressed Data Aggregation) to minimize the medical sensed data. The authors employed compressed sensing with cryptography to compress the data while saving the quality of received data.

The authors in [11–13] presented an adaptive sampling with risk evaluation to decide for monitoring the patients by WBSNs. They proposed a framework to gather medical data by the biosensors and then introduce the risk of the patient using fuzzy logic. Finally, the presented an algorithm for deciding according to the level of patient risk. Shawqi and Idrees [1, 14] introduced a power-aware sampling method using several biosensors to provide the risk of the patient and the best decision to notify the medical experts. First, multisensor sampling based on the weighted scores model is introduced and then they suggested a decision-making algorithm that applied at the coordinator. The works in [15–19] proposed adaptive sampling approaches for WSNs. They have employed similarity measures and some data mining techniques to measure the similarity between two data set of two periods so as to change the sampling rate accordingly. In [20], the authors combine two efficient methods: divide and conquer (D&C) and clustering. They applied the D&C at the sensor nodes and then they applied the enhanced K-means at the cluster node to remove the redundant data and save energy before sending it to the sink. The authors in [6, 7] proposed lossless compression methods for compressing EEG data in IoT networks. In [6], they combine the fractal compression method with differential encoding. In [7], the authors combine Huffman encoding and clustering to further reduce compressed data before sending it to the IoT network.

The authors in [21] proposed three main methods that can be executed at the s-Health architecture, namely, divided in-network processing and resource optimization, event detection and adaptive compression, and dynamic networks association. The first method optimizes medical information transport from the edge node to the healthcare supplier, however taking energy efficiency and application's goodness-of-service request. The second method uses edge computing capabilities to demonstrate an effective data transit architecture that ensures high reliability and rapid emergency reaction. The third method leverages heterogeneous wireless networks within the s-Health architecture to perfect various applications' requirements while optimizing energy consumption and medical data delivery.

The authors in [22] proposed fixing the problem by taking advantage of rapid innovations in the fields of mobile phones, sensors and wireless technology to get better health systems is a critical method. M-Health system accommodates the usage of an edge device to communicate medical data through the wireless network across the m-Health station to diagnose and control the situation of the patient as quickly as possible.

3 EASaM Mechanism

This paper suggests an Energy-saving Adaptive Sampling Mechanism (EASaM) for patient health monitoring in the Internet of Things (IoT) networks. EASaM is executed in the medical biosensors to eliminate unnecessary data during controlling and observing the situation of the patients. It functions in rounds. The round includes two periods. The emergency detection and modifying the sampling rate of each biosensor are two main steps in EASaM. The NEWS is achieved at each biosensor to drop the repeated data before delivering it to the gateway. The sampling rate is adjusted after every two periods based on the situation of the patient at the end of each round. Figure 1 shows the Wireless Body Sensor Network model that used in this paper.

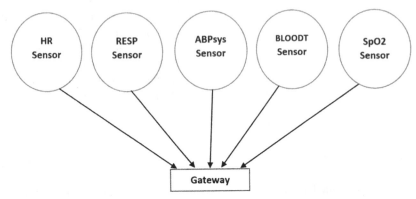

Heart rate (HR), Respiration rate (RESP), Systolic blood pressure (ABPsys), Blood temperature (BLOODT), and Oxygen saturation (SpO2)

Fig. 1. Wireless body sensor network model.

3.1   National Early Warning Score

The medical staff in the hospital utilizes a physiological scoring system named National Early Warning Score (NEWS) to check the situation of the patients to provide the appropriate medical attention and the proper care for the cases with high levels of risk. There are six physiological parameters included in the NEWS that exhibit this system of scoring such as respiratory rate, oxygen saturation, temperature, systolic blood pressure, pulse rate, and awareness level or new contingency [23]. The main feature of NEWS is its simplicity in determining the risk level of the patient using the suitable scores for each type of biosensor. By scoring the sensed values of these biosensors, the NEWS can determine the status of the patient [24]. EASaM mechanism will apply NEWS at each biosensor. NEWS is displayed in Table 1 [25].

Table 1. NEWS (National Early Warning Score).

Physiological parameters	3	2	1	0	1	2	3
Respiration rate	<=8		9–11	12–20		21–24	>25
Oxygen saturation	<=91	92–93	94–95	>=96			
Any supplemental oxygen		Yes		No			
Temperature	<=35		35.1–36.0	36.1–38.0	38.1–39.0	>=39.1	
Systolic BP	<=90	91–100	101–110	111–219			>=220
Heart Rate	<=40		41–50	51–90	91–110	111–130	>=131
Level of Consciousness				A			V, P, or U

3.2 Emergency Detection of the Patient

In medical applications, the measurements of the patients are sensed using several biosensors (e.g., Oxygen Saturation, Heart Rate, Respiration Rate, etc.) located on the body of the patients. These medical sensors send in the periodic way each sensed measurement to the gateway. The gateway receives huge sensed measurements in each period. Hence, the data reduction at each medical sensor is essential before transmitting them to the gateway. Applying data reduction at the biosensors can save energy and extend the lifetime of the monitoring system. Moreover, it can reduce the volume of received data at the gateway to facilitate the analysis to provide an accurate decision about the situation of the patient.

According to NEWS, the medical sensors send to the medical staff just the measurements with scores larger than 0. The measurements of the normal state of the patients will not be transmitted to the gateway. It is clear that periodic monitoring for the situation of the patient will be reduced as well as the transmitted measurements to the gateway are decreased. Finding the relations among the sensed measurements per period before forwarding them to the gateway can participate in solving this problem.

This paper applied the algorithm of emergency detection for the patients at every medical sensor inspirited from [14] with some adjustments to test the scores of sensed measures and forwarding the ones with scores higher than 0. The emergency detection approach is presented in Algorithm 1.

Algorithm 1 Emergency Detection of Patient
Input: MR: Gathered measures in one period
Output: FM: Forwarded measures, SR: scores for forwarded measures
1: $P\_s \leftarrow$ NEWS(MR1)
2: FM $\leftarrow$ FM $\cup$ MR1
3: SR $\leftarrow$ SR $\cup$ P_s
4: ForwardToGateway(MR1)
5: EnergyUpdateForBiosensor()
6: For each sensed measure MRi $\in$ FM do // i = 2,3, ..., N
7:      $C\_s \leftarrow$ NEWS(MRi)
8:      If $C\_s \neq P\_s$ then
9:          SR $\leftarrow$ SR $\cup$ C_s
10:         FM $\leftarrow$ FM $\cup$ MRi
11:         ForwardToGateway(MRi)
12:         EnergyUpdateForBiosensor()
13:         P_s $\leftarrow$ C_s
14:     endif
15: endfor
16: return FM, SR

To further understand the emergency detection technique, an illustrative example will be presented. Suppose there are ten measures of Respiration Rate medical sensor and the period size T = 10, MR = [16, 16, 14, 10, 11, 11, 21, 22, 24, 25]. By using

NEWS, the score's vector of MR measures is SR = [0; 0; 0; 1; 1; 1; 2; 2; 2; 3]. The forwarded measures by the medical sensor are [9, 15, 20, 24]. Only the critical measures and the first measure (even if it was 0) are forwarded to the medical expertise by the biosensor.

3.3   Adapting Sensing Frequency

The accumulated sensed measures of every medical sensor are time-correlated according to the situation of the patient. Consequently, when the patient's situation is stable, a lot of measures would be transferred to the gateway. There are three levels of risk for the patient: (1) low risk represents the normal case of the patient that wants low care by the medical staff. (2) Medium risk represents the middle case between the critical and normal situation of the patient that wants high care by the medical staff, and (3) high risk represents the severe illness cases of the patient that needs consecutive monitoring.

The sensed measures by the medical sensors are time-correlated especially in the cases of high or low risk. Therefore, sending a large volume of sensed measures by the medical sensors leads to spending the energy and increase the load on the medical staff. To get rid of this problem, it is possible to adjust the sampling rate of the medical sensor during sensing the measurements and according to the situation of the patient.

The proposed sampling algorithm in EASaM operates in the way of rounds, where each round includes two periods. Hence, the similarity rate should be calculated by the sampling algorithm between the scores of the measures of the two periods. The edit distance similarity measure is used to calculate this similarity between the two periods. Edit Distance (ED) is a measure of similarity that calculate the distance between two strings. It is also named Levenshtein distance [26]. Algorithm 2 shows the dynamic programming algorithm for computing the edit distance between two sets of data.

Algorithm 2 Edit Distance
Require: Mset1, Mset2: measures for two sets of 2 periods.
Ensure: Mdp: distance between the Mset1 and Mset2.
1: Mdp $\leftarrow$ 0
2: For i $\in$ Len(Mset1)+1 do
3:      For j $\in$ Len(Mset2)+1 do
4:          if i = 0 then
5:              $Mdp_{i,j} \leftarrow j$
6:          else if j = 0 then
7:              $Mdp_{ij} \leftarrow i$
8:          else if $Mset_{i-1}$ = $Mset_{j-1}$ then
9:              $Mdp_{i,j} \leftarrow Mdp_{i-1,j-1}$
10:         else
    11:              $Mdp_{i,j} \leftarrow 1 + \min(Mdp_{i,j-1}, Mdp_{i-1,j}, Mdp_{i-1,j-1})$
12:         endif
13:      endfor
15: endfor
16: return $Mdp_{Len(Mset1), Len(Mset2)}$

Function Len(x) return the length of the set x. The time complexity of the edit distance is Θ(Len(Mset1), Len(Mset2)) and the storage complexity is Θ(Len(Mset1), Len(Mset2)) and this can be improved to Θ(min(Len(Mset1), Len(Mset2))) by observing that the algorithm at any instant requires two columns (or two rows) in the memory storage. Algorithm 3 presents the adaptive sensing rate achieved at every medical sensor at the end of the round.

Algorithm 3 Adaptive Sensing rate Algorithm

Input: SM1, SM2: two measures' sets for 2 periods), ASmin: minimum sensing rate, ASmax: maximum sensing rate

Output: ASrate: new sensing rate

1: Dist ← Edit Distance (SM1, SM2)

2: Dist ← Length(SM1) - Dist

3: SimR ← Dist/ ASmax

4: APsamp ← (1 - SimR) * 100

5: If APsamp < ASmin then

6:          ASrate ← ASmin

7: Else

8:          ASrate ← (ASmax * APsamp)/100

9: endif

10: return ASrate

4   Simulation Results

The performance evaluation of the proposed EASaM approach is introduced in this section. The simulation results are conducted using a custom simulator based on the Python programming language. Real medical data are used during the simulation which is taken from the dataset named MIMIC (Multiple Intelligent Monitoring in Intensive Care) of PhysioNet [27]. EASaM is evaluated using some performance measures like energy consumption, the adaptation of sensing rate vs data reduction, and data integrity. To show the performance of the proposed EASaM, we achieve the comparison with the modified LED* [13]. The simulation is performed during 70 periods (two hours) where the period length is 100 s. The ASmin and ASmax are respectively 10 and 50 measures per period. The record 267n of the patient was used by EASaM during the simulation. The respiration rate medical sensor is used taking into account both high and low risks circumstances.

4.1   Data Reduction vs Adaptation of Sensing Rate

This experiment studied the sensing (sampling) rate modification of the medical sensor and the reduction in the transferred medical data. Figure 2 exhibits the collected results of the suggested EASaM and for two kinds of patients: normal (a) and critical (b) which are compared with modified LED* for the same kinds of patients (c) and (d).

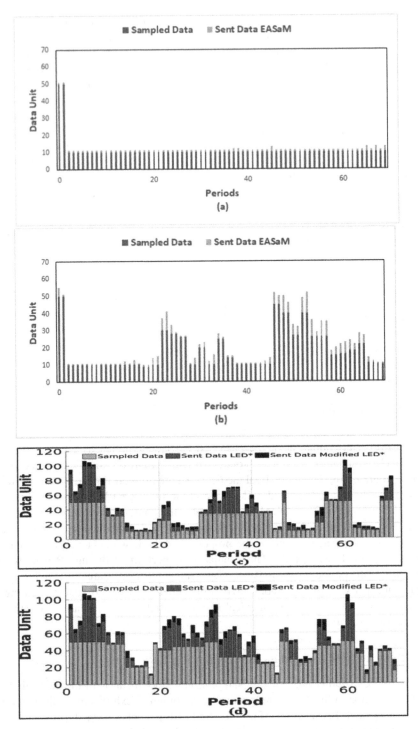

Fig. 2. Adaptation of sensing rate vs data reduction: a) low-risk patient, (b) high-risk patient of EASeT, (c) low-risk patient, and (d) high-risk patient of the modified LED* [13]. (Color figure online)

In Fig. 2(a) and (b), the orange and blue colours denote the number of sent medical data and the number of sampled medical data and respectively after executing the EASaM. In Fig. 2(c) and (d), the light grey and black colours describe the size of medical data and the number of sent data respectively after implementing the EASaM approach. The dark grey colour in Fig. 2(c and d) was ignored which represents the original LED method.

Figure 2(a and b) demonstrates that the size of medical data is altered to a minimum because of the similarity between the scores values of the medical data of the two periods in both cases of the patient: normal and critical. Furthermore, EASaM reduces the size of medical data using the emergency detection strategy by dropping similar data in each period before transferring it to the gateway for both cases of patients: normal and critical. Figure 2(b) shows that the transferred medical data is higher than the transferred data in Fig. 3(a) because it represents the case of the high risk. It can be seen from Fig. 2(a and b) that EASaM has greater performance than modified LED* (c and d) due to minimizing the size of data sent to the gateway and adjusting the rate of sampling of the medical sensor to the minimum.

4.2   Energy Consumption

The energy consumption at the medical sensor is studied taking into account the status of the patient (see Fig. 3). In this study, the EASaM used the same initial energy of the modified LED*, where the initial power of the medical sensor is initialized to 700 units. The consumed energy at the medical sensor for one sensed data is 0.1 and 1 respectively.

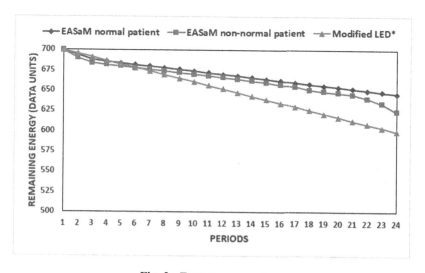

Fig. 3. Energy consumption.

The size of sensed and transmitted medical data has an important impact on the consumed energy at the medical sensor. EASaM approach will highly outperform the modified LED* by reducing the consumed energy at the medical sensor due to reducing the sensed and sent medical data to the gateway.

4.3   Data Integrity

This section studies the effect of the proposed EASaM on data integrity. EASaM presents the results for two cases: normal and critical patients (see Fig. 4a and b), while the results of the modified LED* for a normal patient presented in Fig. 4c.

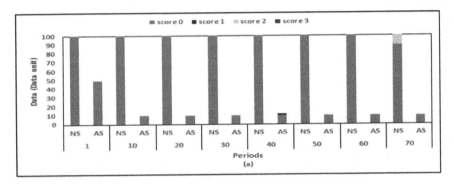

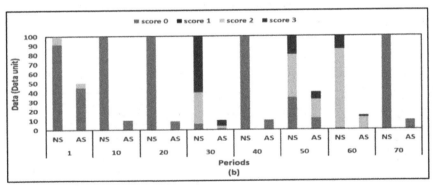

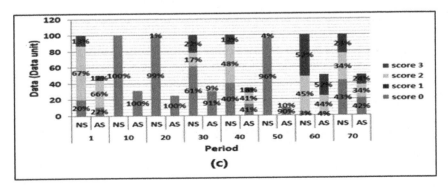

Fig. 4. Data integrity: (a) low-risk patient, (b) high-risk patient of EΛSaM, (c) low-risk patient of the modified LED* [13].

This experiment is performed with Adaptive Sampling (AS) and without Adaptive Sampling (NS) based on the captured medical data per period. It is obtained during scores distribution comparison (NEWS). The scores distribution for NS and AS is achieved for 8 elected periods of 70 periods to show the size of sensed medical data and their scores distributions at the medical sensor. The EASaM decreases the captured medical data for a normal patient in these selected 8 periods up to 85% (see Fig. 4a) compared with modified LED* (see Fig. 4c) that decreases the same captured medical data for the same patient up to 64.5%.

In the normal case of the patient, the kinds of scores are restricted to the score 0. This led to a large decrease in the captured medical data of the medical sensor. Furthermore, the EASaM decreases the medical data for the elected 8 periods up to 80.75% while keeping a proper description of whole scores at the gateway. Therefore, it can be noticed that EASaM guarantees a suitable level of data integrity of the captured medical data whilst preserving all scores without waste at the gateway.

5   Conclusion and Future Work

This paper proposed an energy-saving Adaptive Sampling Mechanism (EASaM) for patient health monitoring in IoT networks. EASaM achieves two main steps: emergency detector and adaptive sampling at each medical sensor. The conducted results show that EASaM is better than modified LED* in terms of energy-saving, data reduction, and data integrity. We plan in the future work to extend the work to achieve multisensor sampling using machine learning and decision support at the gateway to decide the status of the patients.

References

1. Jaber, A.S., Idrees, A.K.: Energy-saving multisensor data sampling and fusion with decision-making for monitoring health risk using WBSNs. Softw. Practice Exp. **51**(2, 271–293 (2021)
2. Harb, H., Mansour, A., Nasser, A., Cruz, E.M., de la Torre Diez, I.: A sensor-based data analytics for patient monitoring in connected healthcare applications. IEEE Sensors J. **21**(2), 974–984 (2020)
3. Dey, N., Ashour, A.S., Bhatt, C.: Internet of things driven connected healthcare. In: Internet of Things and Big Data Technologies for Next Generation Healthcare, pp. 3–12. Springer (2017). Doi: https://doi.org/10.1007/978-3-319-49736-5_1
4. Singh, P.: Internet of things-based health monitoring system: opportunities and challenges. Int. J. Adv. Res. Comput. Sci. **9**(1) (2018)
5. Vitabile, S., et al.: Medical data processing and analysis for remote health and activities monitoring. In: High-Performance Modelling and Simulation for Big Data Applications, pp. 186–220. Springer, Cham, (2019). Doi: https://doi.org/10.1007/978-3-030-16272-6_7
6. Idrees, S.K., Idrees, A.K.: New fog computing enabled lossless EEG data compression scheme in IoT networks. J. Ambient Intell. Humanized Comput., 1–14 (2021)
7. Al-Nassrawy, K.K., Al-Shammary, D., Idrees, A.K.: High performance fractal compression for EEG health network traffic. Procedia Comput. Sci. **167**, 1240–1249 (2020)

8. Idrees, A.K., Jaoude, C.A., Al-Qurabat, A.K.M.: Data reduction and cleaning approach for energy-saving in wireless sensors networks of IoT. In: 2020 16th International Conference on Wireless and Mobile Computing, Networking and Communications (WiMob) (50308). IEEE (2020)

9. Idrees, A.K., Al-Qurabat, A.K.M.: Energy-efficient data transmission and aggregation protocol in periodic sensor networks based fog computing. J. Network Syst. Manage. **29**(1), 1–24 (2021)

10. Soufiene, B.O., Bahattab, A.A., Trad, A., Youssef, H.: Lightweight and confidential data aggregation in healthcare wireless sensor networks. Trans. Emerging Telecommun. Technol. **27**(4), 576–588 (2016)

11. Habib, C., Makhoul, A., Darazi, R., Salim, C.: Self-adaptive data collection and fusion for health monitoring based on body sensor networks. IEEE Trans. Industr. Inf. **12**(6), 2342–2352 (2016)

12. Habib, C., Carol, A.M., Darazi, R., Couturier, R.: Real-time sampling rate adaptation based on continuous risk level evaluation in wireless body sensor networks. In: 2017 IEEE 13th International Conference on Wireless and Mobile Computing, Networking and Communications (WiMob), pp. 1–8. IEEE (2017)

13. Habib, C., Makhoul, A., Darazi, R., Couturier, R.: Health risk assessment and decision-making for patient monitoring and decision-support using wireless body sensor networks. Inf. Fusion **47**, 10–22 (2019)

14. Shawqi Jaber, A., Idrees, A.K.: Adaptive rate energy-saving data collecting technique for health monitoring in wireless body sensor networks. Int. J. Commun. Syst. **33**(17), e4589 (2020)

15. Idrees, A.K., Harb, H., Jaber, A., Zahwe, O., Taam, M.A.: Adaptive distributed energy-saving data gathering technique for wireless sensor networks. In: 2017 IEEE 13th International Conference on Wireless and Mobile Computing, Networking and Communications (WiMob), pp. 55–62. IEEE (2017)

16. Al-Qurabat, A.K.M., Idrees, A.K.: Energy-efficient adaptive distributed data collection method for periodic sensor networks. Int. J. Internet Technol. Secured Trans. **8**(3), 297–335 (2018)

17. Idrees, A.K., Al-Qurabat, A.K.M.: Distributed adaptive data collection protocol for improving lifetime in periodic sensor networks. IAENG Int. J. Comput. Sci. **44**(3) (2017)

18. Al-Qurabat, A.K.M., Idrees, A.K.: Data gathering and aggregation with selective transmission technique to optimize the lifetime of Internet of Things networks. Int. J. Commun. Syst. **33**(11), e4408 (2020)

19. Harb, H., Makhoul, A., Jaber, A., Tawil, R., Bazzi, O.: Adaptive data collection approach based on sets similarity function for saving energy in periodic sensor networks. Int. J. Inf. Technol. Manage. **15**(4), 346–363 (2016)

20. Idrees, A.K., Al-Qurabat, A.K.M., Jaoude, C.A., Al-Yaseen, W.L.: Integrated divide and conquer with enhanced k-means technique for energy-saving data aggregation in wireless sensor networks. In: 2019 15th International Wireless Communications & Mobile Computing Conference (IWCMC), pp. 973–978. IEEE, (2019)

21. Abdellatif, A.A., Mohamed, A., Chiasserini, C.F., Erbad, A., Guizani, M.: Edge computing for energy-efficient smart health systems: data and application-specific approaches. In: Energy Efficiency of Medical Devices and Healthcare Applications, pp. 53–67. Academic Press (2020)

22. Al-Marridi, A.Z., Mohamed, A., Erbad, A., Al-Ali, A., Guizani, M.: Efficient eeg mobile edge computing and optimal resource allocation for smart health applications. In: 2019 15th International Wireless Communications & Mobile Computing Conference (IWCMC), pp. 1261–1266. IEEE (2019)

23. Mukhal, A., Burns, J.M., Raj, R., Sandhu, G.: Implementing the national early warning score (news) for identification of deteriorating patients and measuring adherence to protocol. Eur. J. Internal Med. **24**, e267 (2013)
24. Schein, R.M.H., Hazday, N., Pena, M., Ruben, B.H., Sprung, C.L.: Clinical antecedents to in-hospital cardiopulmonary arrest. Chest **98**(6), 1388–1392 (1990)
25. National Early Warning Score (NEWS), Royal College of Physicians, London, U.K. (2015). http://www.rcplondon.ac.uk/resources/nationalearly-warning-score-news
26. Navarro, G.: A guided tour to approximate string matching. ACM Comput. Surv. (CSUR) **33**(1), 31–88 (2001)
27. Goldberger, A.L., Amaral, L.A.N., Glass, L., et al.: Physiobank, physiotoolkit, and physionet: components of a new research resource for complex physiologic signals. Circulation **101**(23), e215–e220 (2000)

ETOP: Energy-Efficient Transmission Optimization Protocol in Sensor Networks of IoT

Ali Kadhum Idrees[1]([⊠]) [ID], Safaa O. Al-Mamory[2] [ID],
Sara Kadhum Idrees[1] [ID], and Raphael Couturier[3] [ID]

[1] Department of Computer Science, University of Babylon, Babylon, Iraq
ali.idrees@uobabylon.edu.iq
[2] College of Business Informatics, University of Information Technology
and Communications, Baghdad, Iraq
salmamory@uoitc.edu.iq
[3] FEMTO-ST Institute/CNRS, The DISC Department,
Univ. Bourgogne Franche-Comte, Belfort, France
raphael.couturier@univ-fcomte.fr

Abstract. In the past few years, smart devices have been rapidly increased due to their ever-increasing use in different real-world applications. Most of these devices are sensor nodes that represents the basic element in the Internet of Things (IoT). This increasing number in sensor devices will lead to an increase in the size of transmitted sensed readings across the internet, spending energy of sensor nodes, and decreasing the lifetime of the network. Therefore, to tackle this problem, an Energy-efficient Transmission Optimization Protocol (ETOP) is proposed to optimize the transmission and the lifetime of Sensor Networks of IoT. ETOP achieves this mission by using a simple reduction algorithm-based correlation clustering at the sensor stage to remove the redundant data before transmitting it to the gateway or sink. The results are conducted using the OMNeT++ simulator which show that the ETOP protocol can optimize the transmission and the lifetime of Sensor networks better than other methods.

Keywords: IoT · Sensor networks · Transmission Optimization · Data reduction · Energy efficiency

1 Introduction

Smart sensor nodes in the IoT network are used in various applications ranging from health care, military, smart home, safety, agriculture, smart transportation, and so on [1]. These smart devices will keep on rapidly increasing in the near future. These sensor devices represent the basic element in the IoT and the biggest data generator on the IoT network [2]. Therefore, the huge number of IoT sensor nodes leads to massive traffic that must be transmitted over the IoT network. A great part of this transmitted data by the sensor nodes are temporally correlated or similar data. Since the smart sensor devices have limited memory, energy, processing, and communication and since the sending and receiving aspects of the traffic are the ones spending the most energy

© Springer Nature Switzerland AG 2021
A. M. Al-Bakry et al. (Eds.): NTICT 2021, CCIS 1511, pp. 176–186, 2021.
https://doi.org/10.1007/978-3-030-93417-0_12

[3–5], it is therefore necessary to process this data and remove the unnecessary data before transmitting it to the next destination [6]. It is important to design energy-efficient protocols that reduce similar correlated data to save power, thus enhancing the lifetime while saving the data integrity at a suitable level [7]. The contribution of this paper is summarized as follows.

I. This paper suggests an Energy-saving Transmission Optimization Protocol (ETOP) to optimize the transmission and the lifetime of Sensor Networks of IoT. It is implemented in a distributed way at each sensor node and it achieves data collection, data processing and then transmission. During the collection, the sensor gathers the sensed readings and saves them a period of time. After that, the sensor applies a simple reduction algorithm-based correlation clustering to remove the redundant data before transmitting it to the gateway or sink.

II. Some experiments are performed using OMNeT++ simulator [8] and based on real reading from the sensor nodes deployed in the Intel Lab [9]. The simulation results explain the that the proposed ETOP protocol outperforms the other methods PFF [10] and ATP [11] in terms of energy consumption, data accuracy, and data reduction ratio.

The rest of the paper is arranged as follows. Next Section explains the related works. Section 3 introduces the ETOP protocol in more details. Section 4 introduces the simulation results. Section 5 is assigned to the conclusion and future work.

2   Literature Review

This section introduces several related works using different technique to reduce the data collected and improve the lifetime. In [10], the PFF method is executed in the sensor device and aggregator node. The authors use the Jaccard similarity to reduce redundant data in the sensor node while they employ set similarity at the aggregator node to reduce the redundant sets of data. The authors in [11] present a technique called ATP at the sensor device to decrease the data before transmitting it to the sink. They remove the redundancy at the sensor node and then apply some methods different to reduce the spatially correlated data in the gateway. The work in [12, 13] presents a DaT method for lowering the size of sensed data in WSN. The authors suggest a method to cluster the data using a modified K-nearest neighbour. After that, it forwards only one data from each group. Then, they further reduce the data at the second level of the network using the clustering approach. The authors in [14] suggest a TLDA method to enhance the sensor network lifetime. They apply the time series methods in the sensor and aggregator nodes to diminish the redundant data in the WSN.

The works in [15–17] focus on the data gathering and sampling approaches using different similarity techniques to reduce the redundant data during sensing in the same conditions for long periods. The authors in [18–20] propose several data mining techniques and algorithms to reduce similar data either on the sensor nodes and/or aggregators.

In this paper, an Energy-efficient Transmission Optimization Protocol is introduced to optimize the transmission and the lifetime of Sensor Networks of the IoT. ETOP is used in a distributed way at each sensor node and it achieves the data collection, data processing and then the transmission. During the collection, the sensor gathers the sensed readings and saved them for a certain period of time. After that, the sensor implements a simple reduction algorithm-based correlation clustering to remove the redundant data before transmitting it to the gateway or sink.

3   Network Model

In most sensor networks based IoT, the sensor nodes are sending their sensed data periodically, are called periodic sensor networks. The proposed ETOP protocol is based on this type of network. Each sensor device k captures new reading r_i at a slot of time s. After that, the device k constructs a vector of sensed readings $R^k = \{r_1^k, r_2^k, \ldots, r_N^k\}$ at every period p, where R is the number of readings captured during period p and transmits them to a certain gateway or node leader [22, 23]. Figure 1 exhibits an example of periodic sensor nodes where each sensor device captures one reading every five minutes where s = 5 min, and transfers its collected vector of data which includes five readings where N = 5, to the gateway at the end of each period.

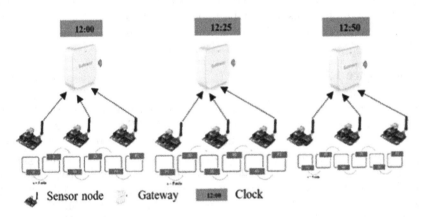

Fig. 1. Periodic sensor network model with illustrative example.

In the periodic data sensing model, it is important to take into account the dynamic changing of the conditions of the monitored area of interest. These conditions can either speed up or slow according to the situation of the surrounding environment [24, 25]. It is expected that the sensor device captures similar or close readings many times especially when the time slot period s has decreased. This can make the sensor device transmit large redundant data during every period. These huge collected readings can highly participate in reducing the energy of the sensor battery. Therefore, it is important to remove the redundant data before forwarding them to the gateway [26].

4   ETOP Protocol

This section illustrates the ETOP protocol for sensor networks of IoT. The lifetime of the network in this protocol is divided into periods and each period includes sensing the reading, processing it, and finally transmitting it to the gateway. The sensor transmission reduction algorithm is implemented to remove the readings redundancy and to decrease the traffic transmission to the next level of the network (sink or gateway). Figure 2 shows the ETOP protocol.

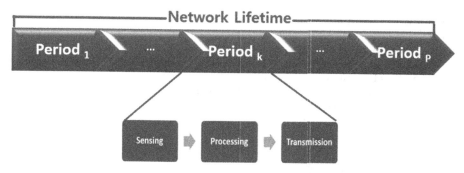

Fig. 2. ETOP protocol.

Each sensor at this stage senses its surrounding environment and sends the sensed readings periodically to the gateway. Therefore, during one period the sensor device collects a vector of readings R = $\{r_1, r_2,..., r_N\}$, where N is the number of readings during one period. The gathered data are similar or very close to each other due to the slow change in the environmental conditions. This temporal correlation can be treated and eliminate the similar reading before transmitting it to the gateway to save energy thus enhancing the lifetime of the network.

Therefore, a simple lightweight reduction algorithm based on correlation clustering algorithm was used to remove the redundant readings to reduce transmitted traffic to the gateway. This saves the energy of sensor nodes and maximizes the life of the batteries of sensor nodes. This algorithm will cluster the readings in the R vector into groups of similar readings, and only one reading for each group will be sent to the gateway. This clustering algorithm is quick and needs one passing over the collected readings R. This algorithm is implemented in a distributed way in each IoT sensor in the network. Algorithm 1 exhibits a Transmission reduction-based correlation clustering algorithm. This algorithm used a different function DiffrenceFunc (see step 8 in Algorithm 1) to allow the close readings to belong to the same group if the difference between them is less than or equal to threshold T. The difference function can be defined as follows

Definition 1 (DiffrenceFunc). We define the difference function between two readings r_i and $r_j \in R$ captured by the sensor device as:

$$\text{DifferenceFunc}(r_i, r_j) = \begin{cases} 1 & if |r_i - r_j| \leq T \\ 0 & otherwise. \end{cases}$$

Algorithm 1 `Transmission reduction-based correlation clustering`

Require: R: set of gathered readings. N: number of readings, T: distance threshold
Ensure: X: reduced data after removing redundancy.
1: $F_i \leftarrow$ false; $// i \leftarrow 1,..., N$
2: $g \leftarrow 1$;
3: $DS_1^g \leftarrow R_1$; $DS_0^g \leftarrow 1$;
4: $F_1 \leftarrow$ true;
5: QuickSort (R) ; // sort the readings in R
6: **for** $k \leftarrow 1$ *to* N **do**
7:     **if** $F_k =$ false **then**
8:         **if** DiffrenceFunc(DS_1^g, R_k, T) = 1 **then**
9:             $DS_0^g \leftarrow DS_0^g + 1$;
10:            $DS_{DS_0^g}^g \leftarrow R_k$
11:        **else**
12:            $g \leftarrow g + 1$;
13:            $DS_0^g \leftarrow R_k$
14:        **endif**
15:            $F_k \leftarrow$ true;
16:    **endif**
17: **endfor**
18: **for** $k \leftarrow 1$ *to* g **do**
19:    $X_k \leftarrow$ *Average* (DS^k, DS_0^k); // average of group DS^k of length DS_0^k
20 : **endfor**
21: **return** X;

In Algorithm 1, the time complexity required to implement this algorithm is O (NlogN). Each sensor node will execute this algorithm at each period to eliminate the redundant readings, save energy, and extend the sensor network's lifetime.

5   Performance Evaluation and Analysis

The proposed ETOP protocol has been evaluated using the OMNeT++ network simulator [8]. Several experiments have been achieved using real readings from sensor devices which are deployed in the Intel Berkeley Lab. [9]. This Lab. includes 47 sensor devices which are responsible for gathering the values of temperature, voltage, humidity, and light every 31 s (see Fig. 3). Table 1 shows the simulation parameters.

Table 1. Parameters values for simulation.

Parameter	Value
N	20, 50, and 100 readings
Nodes number	47 nodes
E_{elec}	50 nJ/bit
B_{amp}	100 pJ/bit/m^2
T	0.03, 0.05, and 0.07

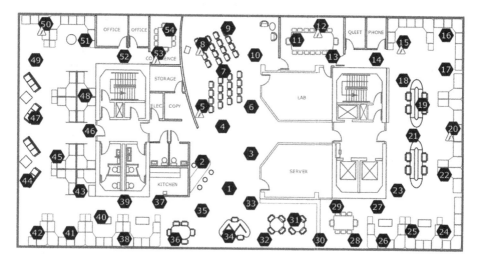

Fig. 3. Intel Berkeley Lab.

The ETOP protocol uses Heintzelman's model [21] for energy consumption, and this model only considers the sending and receiving consumed energy during the simulation (see Fig. 4). The packet size is the number of readings in the period multiplying by 64 bits.

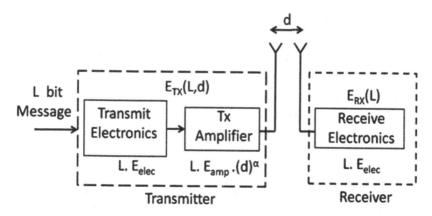

Fig. 4. Heintzelman's model for energy consumption

The proposed ETOP protocol assessed using different performance metrics like energy consumption, transmitted readings, accuracy. It is compared with ATP method [11] and the PFF algorithm [10] results.

5.1 Readings Reduction Ratio

The elimination of unnecessary readings represents one of the essential missions in the IoT sensors. This experiment shows the ratio of data readings after applying the ETOP protocol inside the IoT sensor for Transmission reduction. Figure 5 illustrates the ratio of readings reduction. The ETOP protocol achieves from 94% up to 97% and from 75% up to 81% of transmission reduction by each IoT sensor compared to PFF and ATP respectively. It can be observed that the ETOP protocol decreases the number of transmitted data when N or T increases, because of increasing similar readings (data redundancy) that were removed during each period.

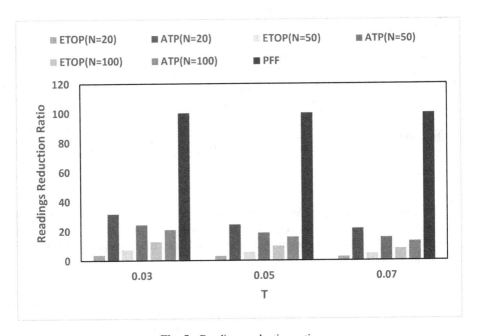

Fig. 5. Readings reduction ratio.

As shown in these results, the ETOP protocol outperforms the other methods in terms of data readings reduction and it decreases the transmission, saves energy, and extends the IoT sensor network lifetime.

5.2 Energy Consumption

Energy saving is one of the most important goals that should be taken into account when designing any protocol for the sensor networks. This section studies the impact of the proposed ETOP protocol on the consumed energy by the sensor nodes. Figure 6 introduces energy consumption by the IoT sensor nodes and uses different sizes of readings. The IoT sensor node reduces the consumed energy by the proposed ETOP protocol from 65% up to 77% and from 58% up to 74% in comparison with PFF and ATP methods respectively. It can be seen that the ETOP protocol increases the amount of saved energy when T or N increases, due to a reduction of transmitted data to the gateway during each period.

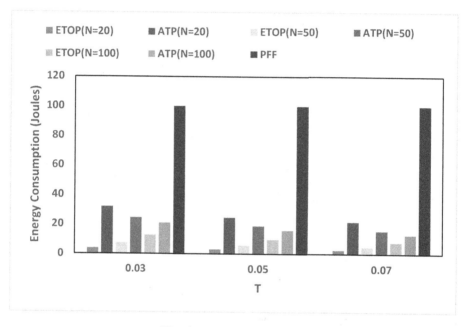

Fig. 6. Energy consumption.

The results show that the ETOP protocol can reduce the consumed power due to a decrease in the transmission traffics at each IoT sensor node before sending it to the next level of the network.

5.3   Accuracy

Since transmission reduction is an essential factor that affects the performance of the network, then it is necessary to reduce the reading traffics, thus reducing the communication cost before transmitting the readings to the next level of the network. At the same time, it is important to keep the quality of the received readings at an acceptable level in the next level of the network. The percentage of reading loss represents the accuracy of received data readings at the next level of the network. Figure 7 presents the percentage of data readings loss for all methods. It can be seen from the conducted results that the proposed ETOP protocol minimizes respectively the percentage of lost data readings from 41% up to 99.8% and from 54% up to 99.8% in comparison with the PFF and the ATP methods.

It can be concluded from the results that the proposed ETOP protocol reduced the transmission volume efficiently while keeping a suitable level of readings accuracy.

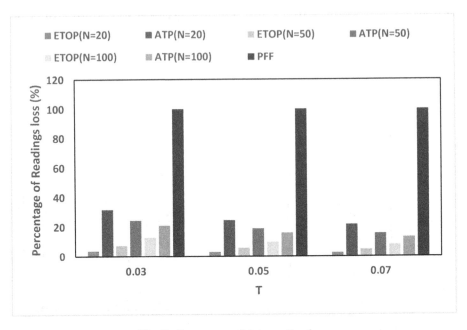

Fig. 7. Percentage of data reading loss.

6   Conclusions and Future Works

This paper proposes an Energy-efficient Transmission Optimization Protocol (ETOP) to optimize the transmission and the lifetime of Sensor Networks of IoT. ETOP accomplishes a simple reduction algorithm-based correlation clustering at the sensor node to remove the redundant data before transmitting it to the gateway or sink. The results show that the proposed ETOP protocol can optimize the transmission of sensor networks better than other methods while maintaining a suitable level of accuracy. In the future, we plan to achieve the transmission reduction on the next level of the network. One could for example, remove the redundant reading resulting from the spatial correlation between the reading of the sensor nodes on the gateway.

References

1. Idrees, A.K., Witwit, A.J.H.: Energy-efficient load-balanced RPL routing protocol for internet of things networks. Int. J. Internet Technol. Secur. Trans. **11**(3), 286–306 (2021)

2. Idrees, S.K., Idrees, A.K.: New fog computing enabled lossless EEG data compression scheme in IoT networks. J. Ambient Intell. Humaniz. Comput. 1–14 (2021). https://doi.org/10.1007/s12652-021-03161-5

3. Idrees, A.K., Deschinkel, K., Salomon, M., Couturier, R.: Multiround distributed lifetime coverage optimization protocol in wireless sensor networks. J. Supercomput. **74**(5), 1949–1972 (2018). https://doi.org/10.1007/s11227-017-2203-7

4. Idrees, A.K., Deschinkel, K., Salomon, M., Couturier, R.: Perimeter-based coverage optimization to improve lifetime in wireless sensor networks. Eng. Optim. **48**(11), 1951–1972 (2016)

5. Idrees, A.K., Al-Yaseen, W.L.: Distributed genetic algorithm for lifetime coverage optimisation in wireless sensor networks. Int. J. Adv. Intell. Paradig. **18**(1), 3–24 (2021)

6. Jaber, A.S., Idrees, A.K.: Energy-saving multisensor data sampling and fusion with decision-making for monitoring health risk using WBSNs. Softw. Pract. Exp. **51**(2), 271–293 (2021)

7. Alhussaini, R., Idrees, A., Salman, M.: Data transmission protocol for reducing the energy consumption in wireless sensor networks. In: Al-mamory, S.O., Alwan, J.K., Hussein, A.D. (eds.) NTICT 2018. CCIS, vol. 938, pp. 35–49. Springer, Cham (2018). https://doi.org/10.1007/978-3-030-01653-1_3

8. Varga, A.: Omnet++. In: Wehrle, K., Güneş, M., Gross, J. (eds.) Modeling and Tools for Network Simulation, pp. 35–59. Springer, Heidelberg (2010). https://doi.org/10.1007/978-3-642-12331-3_3

9. Madden, S.: Intel Berkeley research lab data. Intel corporation 2004 (2003). http://berkeley.intel-research.net/labdata.html. Accessed 08 June 2004

10. Bahi, J.M., Makhoul, A., Medlej, M.: A two tiers data aggregation scheme for periodic sensor networks. Adhoc Sens. Wirel. Netw. **21**(1), 77–100 (2014)

11. Harb, H., Makhoul, A., Couturier, R., Medlej, M.: ATP: an aggregation and transmission protocol for conserving energy in periodic sensor networks. In: Enabling Technologies (2015)

12. Alhussaini, R., Idrees, A., Salman, M.: Data transmission protocol for reducing the energy consumption in wireless sensor networks. In: Al-mamory, S.O., Alwan, J.K., Hussein, A.D. (eds.) NTICT 2018. CCIS, vol. 938, pp. 35–49. Springer, Cham (2018). https://doi.org/10.1007/978-3-030-01653-1_3

13. Idrees, A.K., Alhussaini, R., Salman, M.A.: Energy-efficient two-layer data transmission reduction protocol in periodic sensor networks of IoTs. Pers. Ubiquit. Comput. 1–20 (2020). https://doi.org/10.1007/s00779-020-01384-5

14. Al-Qurabat, A.K.M., Idrees, A.K.: Two level data aggregation protocol for prolonging lifetime of periodic sensor networks. Wireless Netw. **25**(6), 3623–3641 (2019). https://doi.org/10.1007/s11276-019-01957-0

15. Idrees, A.K., Harb, H., Jaber, A., Zahwe, O., Taam, M.A.: Adaptive distributed energy-saving data gathering technique for wireless sensor networks. In: 2017 IEEE 13th International Conference on Wireless and Mobile Computing, Networking and Communications (WiMob), pp. 55–62. IEEE (2017)

16. Idrees, A.K., Al-Qurabat, A.K.: Energy-efficient adaptive distributed data collection method for periodic sensor networks. Int. J. Internet Technol. Secur. Trans. **8**(3), 1951–1972 (2017)

17. Idrees, A.K., Al-Qurabat, A.K.M.: Distributed adaptive data collection protocol for improving lifetime in periodic sensor networks. IAENG Int. J. Comput. Sci. **44**(3), 345–357 (2017)

18. Shawqi Jaber, A., Idrees, A.K.: Adaptive rate energy-saving data collecting technique for health monitoring in wireless body sensor networks. Int. J. Commun. Syst. **33**(17), e4589 (2020)

19. Idrees, A.K., et al.: Distributed data aggregation based modified k-means technique for energy conservation in periodic wireless sensor networks. In: 2018 IEEE Middle East and North Africa Communications Conference (MENACOMM). IEEE (2018)
20. Abdulzahra, S.A., Al-Qurabat, A.K.M., Idrees, A.K.: Compression-based data reduction technique for IoT sensor networks. Baghdad Sci. J. **18**(1), 0184–0184 (2021)
21. Idrees, A.K., et al.: Integrated divide and conquer with enhanced k-means technique for energy-saving data aggregation in wireless sensor networks. In: 2019 15th International Wireless Communications & Mobile Computing Conference (IWCMC). IEEE (2019)
22. Heinzelman, W.R., Chandrakasan, A., Balakrishnan, H.: Energy-efficient communication protocol for wireless microsensor networks. In: Proceedings of the 33rd Annual Hawaii International Conference on 2000 System Sciences, p. 10. IEEE (2000)
23. Idrees, A.K., Al-Qurabat, A.K.M.: Energy-efficient data transmission and aggregation protocol in periodic sensor networks-based fog computing. J. Netw. Syst. Manag. **29**(1), 1–24 (2021)
24. Hameed, M., Idrees, A.: Distributed DBSCAN protocol for energy saving in IoT networks. In: Bindhu, V., Tavares, J.M.R.S., Boulogeorgos, A.-A.A., Vuppalapati, C. (eds.) International Conference on Communication, Computing and Electronics Systems. LNEE, vol. 733, pp. 11–24. Springer, Singapore (2021). https://doi.org/10.1007/978-981-33-4909-4_2
25. Idrees, A.K., Couturier, R.: Energy-saving distributed monitoring-based firefly algorithm in wireless sensors networks. J. Supercomput. 1–26 (2021). https://doi.org/10.1007/s11227-021-03944-9
26. Hameed, M.K., Idrees, A.K.: Cuckoo scheduling algorithm for lifetime optimization in sensor networks of IoT. In: Suma, V., Chen, J.-Z., Baig, Z., Wang, H. (eds.) Inventive Systems and Control. LNNS, vol. 204, pp. 171–187. Springer, Singapore (2021). https://doi.org/10.1007/978-981-16-1395-1_14

Applied Computing

The Extent of Awareness of Faculty Members at Albaydha University About the Concept of Educational Technology and Their Attitudes Towards It

Nayef Ali Saleh Al-Abrat[1]
and Mohammed Hasan Ali Al-Abyadh[2,3]([✉]) [iD]

[1] College of Education and Science in Rada'a, Albaydha University,
Rada'a, Albaydah, Yemen
dr.nayefalabrat78@gmail.com
[2] College of Education in Wadi Addawasir,
Prince Sattam bin Abdulaziz University, Alkharj, Saudi Arabia
m.alabyadh@psau.edu.sa
[3] College of Education, Thamar University, Thamar, Yemen

Abstract. The study aimed to find out the awareness of faculty members at Albaydha University about the concept of educational technology and their attitudes towards it. In order to achieve the aim of the study, the researcher followed the descriptive approach. He also designed the study tool (questionnaire) that consisted of (38) items distributed in two areas and was of sufficient validity and reliability for the purposes of this study. The study sample consisted of all (96) faculty members. After collecting the data using the study tool, the researcher analyzed it statistically, and used in this the arithmetic means, standard deviations, percentages, and the (T-test). The results of the study were as follows:

- The results revealed that the extent of the study sample's awareness of the concept of educational technology is low.
- There is statistically significant differences awareness of the study sample members of the concept of educational technology due to the variable years of experience in education, and in favor of the study samples who have less than five years of experience.
- The results reveal that there is an average positive trend in the study sample towards educational technology.
- There are statistically significant differences of the trend towards educational technology due to the years of experience in education variable, and in favor of the study samples who have less than five years of experience.
- There is a positive correlation relationship with statistical significance between the extent of the study sample's awareness of the concept of educational technology and their level of attitudes towards it.

Keywords: Educational technology · Attitude · Faculty members

A. M. Al-Bakry et al. (Eds.): NTICT 2021, CCIS 1511, pp. 189–207, 2021.
https://doi.org/10.1007/978-3-030-93417-0_13

1 Introduction

The tremendous development in human knowledge and experiences has resulted in many aspects and developments in all fields, the most important of which is technology. This technology was initially produced and employed in non-educational fields, such as the industrial, medical, and other fields. Then, this technology was adapted and used in the educational field, and it was called educational technology. Many workers in the field of educational technology have broad hopes for the role that educational technology can play if it is best used and employed in the educational learning process. As educational technology includes devices, tools, materials, educational situations, teaching strategies, evaluation and feedback, it is included in all educational fields, which leads to effective development and a noticeable increase in the outcomes of the educational process (Ashteiuh and Olayan 2010: 37).

Its use in teaching and learning processes works to support education, enhance learning, and provide an opportunity for both the teacher and the learner to use this technology in an appropriate manner and to achieve educational goals. The intelligent and skilful use of appropriate educational technology to teach the learner individually and collectively increases the teacher's level in his work and raises the level of the learner in his learning (Al-Jamal 2004: 124; Labusch and Eickelmann 2018). Despite the great importance attached to educational technology and its role in the teaching and learning process, its concept is still tainted with much ambiguity. There are some who believe that the concept of educational technology narrows to be limited to educational aids, including the devices and tools they contain (Salama 2001: 125).

This common belief towards educational technology is unfair, as it adds a lot of ambiguity to the content of the term and limits its adoption in the fields of educational system development in all its aspects. This explains that technology, for many, means machines and electronic devices that represent tangible aspects of technology that are used in everyday life. They lose sight of the intangible aspects of technology, namely the complex processes, systems and tasks that must be planned, managed, and evaluated in order to obtain the best. It means the orderly application of scientific knowledge, that is, it includes both the theoretical and the practical sides. It provides a cognitive framework to support the application (Shami et al. 2008: 15).

This ambiguity may prevent educational technology from adopting and employing it ineffectively during the teaching process by faculty members. The results of some studies, such as the study of Clark (2000, 179), Sultan (2001, 166), Haji Issa (2001, 215) and Salem (2004, 104) indicate that one of the most important reasons for faculty members' resistance to using educational technology is due to the ignorance of the teaching staff or their lack of awareness and lack of their culture about the concept of educational technology, lack of belief in the scientific value of educational technology, in addition to their poor awareness of the importance and role of its use in education. Salama (2007, 122) adds the teacher's negative attitude towards educational technology, as some faculty members see that educational technology is a competition for them, without being aware of the role of the new teacher, which has changed under educational technology from a mere conveyor of information to a manager of the educational position, providing the necessary facilities for education, a designer for the educational process, a producer of educational materials, a guide for the learner, and a constant evaluation of the educational system. This reflects negatively on the attitudes of faculty members towards educational technology. The results of many studies indicates that the attitudes of faculty members towards educational technology affect their use of it.

1.1   The Problem of the Study

Despite the results of many studies dealing with the issue of educational technology, and what some researchers reported, such as Salem and Saraya (2003: 277), Salem (2004), Salama (2007: 116) and Shammi et al. (2008: 20) the benefits of using educational technology and its importance in supporting the performance of the teacher, saving his effort and time, overcoming the problem of increasing the numbers of learners, facing the increase in scientific knowledge, treating the problem of individual differences between learners, achieving learning in its various cognitive, skill and emotional aspects, and increasing the motivation of learners to learn, participate and pay attention. As well as developing the learner's ability to self-learning, increasing his linguistic wealth, training in sound scientific thinking methods, and solving some of the problems of faculty members.

However, the researcher noticed through his experience in field work as a faculty member that there is negligence or evasion by faculty members in employing educational technology and exploiting its potential in the teaching and learning process, and their reliance largely on the style of lecture and class discussions, as they consider it the best way to transfer skills and information and discussing it with their students.

The researcher felt through this that the reason may be due to the lack of awareness of faculty members of the concept of educational technology and its role in the teaching and learning process.

Accordingly, the issue of awareness of the concept of educational technology and the importance of employing it on the ground is a matter of great importance because of its impact on the development of educational reality. Awareness of the concept of educational technology is an essential step for feeling the importance of employing it in the educational field and adopting it in the field. Moreover, identifying the extent of the faculty members' awareness of the concept of educational technology and their attitudes towards it is a necessary matter that reveals the negative aspects and limits them, enhances the positive aspects, and prompts the suggestion of solutions to get rid of these negative aspects.

Although there are Arab and foreign studies that dealt with such a topic in educational technology, but within the limits of the researcher's knowledge, no study has been conducted on the awareness of faculty members about the concept of educational technology and their attitudes towards it in the Republic of Yemen. Therefore, the researcher felt the need to know the awareness of the faculty members at Albaydha University of the concept of educational technology and their attitudes towards it.

Specifically, the study tried to answer the following questions:

– To what extent are the faculty members at Albaydha University aware of the concept of educational technology?

- Are there statistically significant differences in the awareness of faculty members at Albaydha University with the concept of educational technology due to years of teaching experience (less than 5 years - more than 5 years)?
- What are the attitudes of faculty members at Albaydha University towards educational technology?
- Are there statistically significant differences in the attitudes of faculty members at Albaydha University towards educational technology due to years of teaching experience (less than 5 years - more than 5 years)?
- Is there an overall relationship between the awareness of faculty members at Albaydha University about the concept of educational technology and their attitudes towards it?.

1.2   Importance of the Study

The importance of the study stems from the role of educational technology and the advantage it occupies in the educational-learning process in general, as educational technology, if used optimally, can play an effective role in developing the educational system, and reducing educational problems for all school levels.

The results of this study may reveal ambiguities in the concept of educational technology, and may change the negative perception of its role in the educational process and limit it, reinforce the positive aspects, and push to propose solutions to get rid of these negative points.

1.3   The Aims of the Study

The current study aims to:

- Knowing the extent of awareness of faculty members at Albaydha University about the concept of educational technology.
- Exposing the trends of faculty members at Albaydha University towards educational technology.
- Knowing the effect of the variable years of experience in education on the awareness of faculty members at Albaydha University about the concept of educational technology.
- Knowing the impact of the variable years of experience in education on the level of attitudes of faculty members at Albaydha University towards educational technology.
- Revealing the nature of the relationship between the awareness of faculty members at Albaydha University about the concept of educational technology and their attitudes towards it.

1.4   Limits of the Study

The limits of the current study are as follows:

- Human limits: This study was limited to the (96) faculty members in the College of Education and Science in Radaa - Albaydha University.

- Spatial limits: College of Education and Science in Radaa, Albaydha University.
- Subject limits: The subject of this study was limited to studying the awareness of faculty members at Albaydha University about the concept of educational technology and their attitudes towards it.
- Time limits: This study was conducted in the first semester of the academic year (2020/2021).

1.5 Procedural Definitions

Educational Technology: defined by the American Society for Educational Communication Technology as "theory and practice, the design, development, use, and management of processes and resources for learning" (Ghazzawi 2007: 63).

In this research, it is defined as an integrated organization that includes the tools, devices, materials and educational situations that the teacher uses in order to improve the teaching and learning process.

Awareness: is the level of familiarity of faculty members with an appropriate degree of knowledge of the concept of educational technology and its use in university teaching.

Direction: It is a set of ideas, beliefs and knowledge (the perceptual component), which includes a positive or negative evaluation of feelings, or emotion that does not charge the idea (the emotional component), and thus a state of readiness to action (a behavioral element) is formed.

2 Previous Studies

The two researchers reviewed some studies closely related to the subject of the current study, and they were arranged chronologically, and the most prominent of these studies are the following:

Abdul Majeed (2000) This study aimed to find out the extent of awareness of science teachers in the preparatory stage of the innovations of educational technology and their trends towards its use. To achieve this goal, the researcher prepared a list of the most important innovations in educational technology in the field of science education, and a questionnaire to measure science teachers' awareness of educational technology innovations and a scale to measure their attitudes towards its use. Then, he applied these tools to a sample of (365) middle school teachers in some educational departments in the governorates of Cairo, Giza, Qalyubia and Menou-fia.

The results indicated:

- The low level of awareness of science teachers in the preparatory stage of educational technology developments.
- The existence of a positive relationship between awareness of educational technology innovations and the trend towards its use.

The Al-Kadiri and Shdeifat Study (2002).

The study aimed to determine the level of culture attained in the educational computer of the managers and teachers of the Northern Badia in Jordan, and the extent of its difference for them according to their gender, job type, level of qualification, and teaching experience.

To achieve the aim of the study, the researcher used a tool to measure the level of culture, which consisted of (57) items, and the study sample consisted of (196) individuals, including (77) managers and directors, and (119) teachers.

The results of the study showed:

– The low level of culture obtained in the educational computer among the members of the study sample in general.

The study of Hong et al. (Hong et al. 2003):

This study aimed to find out the university students' attitudes towards using Internet technology as an educational method. The sample of the study consisted of (88) male and female students, who are studying in five colleges at the University of Malaysia. To achieve the goal of the study, the researchers used a scale of seven items to measure their attitudes towards using Internet technology as an educational tool.

The results of the study showed a positive trend towards the use of Internet technology in education, and there were no differences in this trend between the sexes, nor be-tween the high and low in the GPA, while there were statistically significant differences related to the type of college, as the trend increased among students of the Faculties of Engineering, Science and Technology in a significant way for students of the College of Human Development.

Al-Ghishan Study (2005):

This study aimed to find out the degree of interest of teachers of basic education in public schools in the Amman Directorates of Education in educational technology, and the difficulties and problems that teachers and learners face when using educational technology in the field. The study sample consists of two groups, namely all teachers of the tenth grade in public schools in Amman, whose number is (3444) teachers, and all students of the tenth grade in government schools in Amman, whose number is (25390). To achieve the aim of the study, two questionnaires were constructed, one for teachers and the other for students.

The results of the study showed that:

– There is an interest in the study sample in educational technology.
– There is no difference between the study sample individuals in their interest in educational technology due to gender.

Hakami's study (2010):

The study aimed to identify the reality of the culture and use of faculty members in scientific colleges at Umm Al-Qura University for information and communication technology in teaching. The researchers adopted the descriptive and analytical approach, and used the questionnaire as a tool to measure the goal of the study. The study sample consisted of (126) faculty members in the scientific colleges at Umm Al-Qura University. The study concluded with a set of results, the most important of which are:

- The awareness of the teaching staff in the scientific colleges of information and communication technology was moderate.
- The degree of use of information and communication technology by faculty members was moderate.
- There are no statistically significant differences between the opinions of faculty members on the culture and use of information and communication technology according to the variable of sex, type of college, rank, and number of years of experience.

Al-Qahtani and Al-Muaither study (2016):

This study aimed at identifying the awareness of the faculty members at Princess Noura University in the Kingdom of Saudi Arabia with stereoscopic imaging technology (hologram) in distance education by measuring the importance of holograms, the difficulties facing its application and their attitudes towards using this technology in teaching. The study tools were applied to a sample of the faculty members at Princess Nora University, the number of which reached (100) faculty members in all the colleges of the university. The study questionnaire was designed from three axes, and one of the most important results of the study was the approval of the study sample on the im-portance of applying hologram technology in teaching. While there were no statistically significant differences in the attitudes of the sample members about all the study axes according to the difference in the degree variable, the type of college and the number of years of experience, which confirms the awareness of the faculty members of the importance of applying these modern techniques in teaching.

The Sharif Study (2018):

This study aimed to measure and determine the extent of awareness of digital and smart educational technologies for faculty members in Saudi universities and their attitudes towards them. The researcher followed the descriptive approach, and used two research tools: a questionnaire on the extent of awareness of faculty members in Saudi universities about digital and smart educational technologies in education. The number of the study sample reached (15) members of the faculty in three Saudi universities. The research reached several results, the most important of which were: there were no statistically significant differences in the degree of awareness of faculty members in Saudi universities about digital and smart educational technologies due to the basic effect of the degree or gender. The results also found that there are statistically significant differences in the attitudes of the faculty members due to the basic effect of the academic degree.

Al-Shair's study (2020):

This study aimed to reveal the awareness of home economics students about employing the latest educational and information technology and their motivation for achievement. To achieve the aim of the study, the researcher used the descriptive approach. The research community consisted of (1309) male and female students, who were randomly selected, and consisted of (240) male and female students. The researcher also prepared the study tools (questionnaire on educational and information technology innovations and the measure of motivation for achievement) and they were applied to the study sample members. Statistical treatment methods were applied using the (SPSS) program, and the study results demonstrated the study sample's awareness of employing educational and information technology innovations. The results also showed an increase in students' motivation for achievement.

Muhammad's study (2020):
The study aimed to know the extent of awareness of using e-learning in the education of students with special needs in the Dhofar Governorate of the Sultanate of Oman, according to the following indicators: type of disability, degree of disability, educational level, and the gender of the teacher on which the study questions were based. The researcher used the descriptive approach due to its suitability for the purposes of the study. The sample of the study consisted of (50) male and female teachers who are currently teaching students with special needs in the Dhofar Governorate. The study reached a number of results, the most important of which is that the academic preparation does not train teachers sufficiently to use e-learning and then direct it to students with special needs, which led to a decrease in the awareness of teachers who use these methods with students with special needs.

El-Gohary Study (2020):
This study aimed to measure the awareness of faculty members at Prince Sattam bin Abdulaziz University using the e-learning platform in light of the outbreak of the covid 19 virus and their attitudes towards learning by deduction. The researcher used the descriptive and quasi-experimental approach and the statistical demographic variables for their relevance for the purposes of the study. The sample of the study consisted of (100) faculty members at Prince Sattam bin Abdulaziz University, who were chosen in a stratified random manner. The results of the study showed that the awareness of the faculty members at Sattam bin Abdulaziz University about the e-learning platform was high, and their attitudes towards the learning environment in the survey were also high.

General comment on previous studies:
By reviewing what has been presented from previous studies, the two researchers concluded the following:

– Some studies that dealt with the issue of awareness, culture, or the importance of the concept of educational technology revealed contradictions in the results as follows:
– The results of some studies showed low awareness of the study sample members of educational technology, such as the study of (Abdul Majeed (2000), Al-Qadri and Al-Shdeifat (2002). There are also statistically significant differences in the degree of awareness due to the variable of years of experience in favor of fresh graduates and those with short experience. The results of the study of Hakami (2010) and Muhammad (2020) showed the study sample's awareness of educational technology, but with a medium degree, in addition to the absence of differences attributable to the years of experience variable, while the results of some studies showed a high awareness and interest in educational technology, such as the study of both the fog Al-Qahtani and Al-Muaither (2016), Al-Sharif (2018), Al-Saeed (2020) and El Gohary (2020), in addition to the absence of statistically significant differences due to the years of experience variable.

- Some studies revealed positive trends towards educational technology, such as the study Hong et al. (2003), Al-Sharif (2018) and El Gohary (2020), while the trends towards educational technology were neutral in the results of Abdul Majeed (2000) and Al-Saeed (2020).
- Abdul Majeed's study (2000) revealed a positive correlation between awareness of the concept of educational technology and the trend towards it.
- It is evident through the presentation of previous studies that there is a difference in the results reached by some studies, and the researcher did not find, according to his knowledge, any study dealing with this topic in the Republic of Yemen. That is why the two researchers tried through this study to identify the extent of awareness of faculty members at Albaydha University about the concept of educational technology and their attitudes towards it.

3   Methodology and Procedures

Study methodology: The researcher used the descriptive and analytical method because it is one of the most appropriate scientific research methodologies suitable for the nature of this study to answer its questions on the one hand, and to achieve its objectives on the other hand. It is the approach that studies a phenomenon or an existing issue from which information can be obtained that answers the research questions without the interference of the researcher, in order to describe and interpret the results of the research (Al-Agha and Al-Ustaz 2002: 83).

Study population: The study population consists of faculty members at Albaydha University for the academic year 2020–2021, whose number is approximately (190) faculty members. Study sample: The study sample consists of all faculty members in the College of Education and Science in Radaa, who actually carry out the teaching process in the college, whose number is (96) faculty members (Table 1).

Table 1. Distribution of the study sample according to years of teaching experience

Number	Experience (in years)	
	More than 5 years	Less than 5 years
96	64	32
Percentage	67%	33%

Study tool: The researcher prepared the study tool (questionnaire), through the following:

- Review the educational literature related to the subject of study.
- Relying on the opinions and ideas of those with experience in the field of educational technology, curricula and methods of teaching.

In light of what was included in the previous sources, the study tool (the questionnaire) was prepared, and the number of items is (38) organized and distributed on two domains according to the requirements of the current study.

The study tool consists of two parts:

Section One: General information related to years of experience in education.

The second section: It includes the questionnaire domains, which are as follows:

The first domain: It relates to the concept of educational technology and contains (16) items.

The second domain: It relates to the attitudes of faculty members towards educational technology and contains (22) items.

The validity of the study tool (questionnaire) prepared by the researcher was verified by presenting it to a group of juries with specialization in the field of educational technology, curricula and teaching methods. They were asked to express their opinion on the study tool in terms of the linguistic formulation, the extent of clarity and scientific accuracy of each item, the extent to which the paragraphs belong to the domain they fall under, as well as the extent of the comprehensiveness of each domain of the questionnaire, and to make any observations they deem appropriate.

Based on the opinions of the juries and their observations, the linguistic wording of some items was modified, some items were deleted, and new ones were added. The items that got (80%) from the juries' consensus were kept as appropriate items to measure the goals for which they were set.

The stability of the study tool (questionnaire) prepared by the researcher was also confirmed by applying it to a pilot sample consisting of (15) teachers. Then, the same study tool (the questionnaire) was applied again to the same sample two weeks after it was applied for the first time. The correlation between the two applications was calculated using Pearson's coefficient, with a correlation coefficient of (0.86). This ratio is considered an acceptable stability factor for the purposes of this study. The internal consistency factor of the stability of the paragraphs of the resolution was also calculated using the Cronbach Alpha equation, with a value of (0.88). These reliability coefficients were considered sufficient for the purposes of this study. After making sure of the validity and reliability of the study tool, the researcher applied it to the study sample.

Statistical Analysis:

To answer the study questions, arithmetic means, standard deviations, percentages, and T-test in addition to the Pearson correlation coefficient were used wherever necessary. The researcher adopted the pentagonal gradient to know the extent of the study sample's awareness of the concept of educational technology and the direction towards it, and this gradient was classified according to the arithmetic averages, as shown in the following Table 2:

Table 2. Classification of scores within the arithmetic means

The numerical category of the arithmetic mean	Value
1–1.80	Strongly agree
1.81–2.60	Agree
2.61–3.40	Neutral
3.41–4.20	Disagree
4.21–5	Strongly disagree

4 Results

Results related to the first question, which stated: To what extent are faculty members at Al Baydha University aware of the concept of educational technology?

To answer this question, the researcher calculated the arithmetic averages and T-test for one sample, as well as the standard deviation to find out the degree of consensus or disparity of opinions among the members of the research sample towards each of the items of the scale of awareness of the concept of educational technology, as shown in the following table:

Table 3. The averages, deviations, percentages, and "c" values for the items of the Awareness Scale

#	Item	Arth. mean	Std. Dev.	"T" value	Sig.	Mean	Materiality	Level to mean
			0.50	14.370	0.00	0.93	49.40	Low
			0.52	13.124	0.00	0.85	51.00	Low
3	Educational technology is the same as educational aids	1	Seeing educational technology as an effective means of transmitting information	2.47	0.00		0.65	Average
4	Think of educational technology as a subject in itself	2	Teaching aids are part of the educational technology system	2.55	0.00	0.97	48.60	Low
5	No matter how advanced educational technology can be, it will not be a catalyst in education	2.73	0.66	7.819	0.00	0.67	54.60	Average
6	Consider pre-made educational programs and materials an integral part of educational technology	2.53	0.51	13.344	0.00	0.87	50.60	Low
7	Educational technology is nothing but a set of educational devices that can be used in education	2.78	0.61	7.790	0.00	0.62	55.60	Average
8	Educational technology is concerned with the practical aspects only	2.60	0.53	11.754	0.00	0.80	52.00	Low
9	I agree to join any training course that helps in training in the use of technology in education	2.67	0.57	9.923	0.00	0.73	53.40	Average
10	I am able to teach what I want without the use of educational technology	2.48	0.50	14.090	0.00	0.92	49.60	Low
11	I see that educational technology helps solve many educational problems	2.50	0.53	13.826	0.00	0.90	50.00	Low
12	I think it is inappropriate to waste my lecture time showing an educational film that is directly related to the subject I am studying	2.58	0.56	12.724	0.00	0.82	51.60	Low

(continued)

Table 3. (*continued*)

#	Item	Arth. mean	Std. Dev.	"T" value	Sig.	Mean	Materiality	Level to mean
			0.50	14.370	0.00	0.93	49.40	Low
			0.52	13.124	0.00	0.85	51.00	Low
13	I want to increase my knowledge about the concept of educational technology and its educational applications	2.70	0.67	8.075	0.00	0.70	54.00	Average
14	I see educational technology as everything that influences the educational situation	2.73	0.58	10.000	0.00	0.67	54.60	Average
15	Educational technology means the orderly application of scientific knowledge	2.38	0.52	15.036	0.00	0.102	47.60	Low
16	I use educational technology because of its importance in expanding students' perceptions during education	2.63	0.52	11.428	0.00	0.77	52.60	Average
Total Score		2.59	0.56	2.800	0.00	0.81	51.80	Low

It is evident from the data of Table 3 that the arithmetic averages in general for the items of the level of Awareness Scale came at low rates, as the total arithmetic mean of the items of the scale was (2.59), with an approval of the content of the items that measure the level of awareness (51.80%). This confirms that the extent of the study sample's awareness of the concept of educational technology was low.

Item No. (7) related to (educational technology is nothing but a group of educational devices that can be used in education) ranked first with an average arithmetic mean of (2.78), and with an approval rate of (55.60%), on the content of this paragraph. Item (3), pertaining to (educational technology is the same as educational aids) came in second place, with an average arithmetic mean of (2.75), and with an approval rate of (55.00%) on the content of this item. Item (15) came in last place related to (educational technology means the orderly application of scientific knowledge), with a low arithmetic average of (2.38), and with an approval rate of (47.60%) for the content of this item. As for the most consistent items in the level of awareness of the concept of educational technology among the study sample, items (1 and 10) were the most consistent were the standard deviation for each of them (0.50), followed by item (6) with a standard deviation (0.51).

The results of (t) test for one sample (One-Sample Test) came in support of what was reviewed, where the average difference between the items of the awareness scale and the average approved (3.40) ranged between (0.62) and (0.102). The values of (t) at this level of difference were (7.790) and (15.036) respectively, and all of them are statistically significant values at the level of significance (0.05).

In general, the average difference between the mean of the total awareness level and the approved average value of the consciousness measure was (0.81). The value of (t) at this level of difference was (2.800), which is a statistically significant value at the level of significance (0.05). This means that the awareness of the study sample individuals about the concept of educational technology is low.

Accordingly, it can be said that the concept of educational technology is still confused and ambiguous among the faculty members. The deficient view of educational technology as meaning educational aids only, and that it is nothing but a group of educational devices that can be used in teaching in classrooms, leading to their ineffective use, and this may reflect negatively on the faculty members adopting them in actual teaching.

This may be due to the lack of training courses for faculty members in the field of educational technology to clarify its concept, and the potentials that it can offer in developing the teaching and learning process and upgrading the performance of faculty members, if they are best used in university teaching. In addition to the low level of training programs for faculty members as these programs are traditional and distinguished by the supremacy of theoretical aspects over the process, and there is no indication of training programs in the field of educational technology. This result is consistent with the results of the study of Abdul Majeed (2000), Al-Qadri and Al-Shdeifat (2002), which showed the low level of understanding and culture of the study sample in the concept of educational technology. While the results of this study differed with the results of Hakami (2010), Muhammad (2020), Al-Gheishan (2005) Al-Qahtani and Al-Muaither (2016), Al-Sharif (2018), Al-Shaer (2020), Al-Saeed (2020) and El Gohary (2020) that showed a good, acceptable or average level of awareness of the concept of educational technology.

- Results related to the second question, which stated: Are there statistically significant differences in the awareness of faculty members at Albaydha University with the concept of educational technology attributable to years of teaching experience (less than 5 years - more than 5 years)?

To answer this question, arithmetic averages and standard deviations were calculated for the extent of the study sample's awareness of the concept of educational technology, according to the years of experience variable (less than 5 years - more than 5 years), and the (t) test was used to find out the significance of the differences between the arithmetic averages. And the results were as shown in the following table:

Table 4. Arithmetic averages, standard deviations and the value of (t) test for differences between the averages of the study sample's awareness of the concept of educational technology according to the years of experience in education variable

#	Teaching experience variable	No.	Mean	Sig.	T value	Level of significance
1	Less than 5 years	23	45.52	4.42	4.178	0.000
2	More than 5 years	37	39.05	6.55		

It is evident from the results of Table 4 that there is an effect of the variable years of experience in education on the level of awareness of the concept of educational technology, where the total arithmetic average of the awareness of the study sample who have less than five years' experience reached (45.52), with a standard deviation of

(4.42). While the total arithmetic mean of the awareness of the study sample who have more than five years' experience was (39.05), with a standard deviation of (6.55). The value of "T" was (4.178), with a significant level of (0.000). This indicates the existence of statistically significant differences at the level of significance ($\alpha = 0.05$) between the average awareness level and in favor of the study sample who had less than five years of experience. This means that faculty members who are new to education have an awareness of the concept of educational technology, more than faculty members who have been in education for five years or more.

The researcher attributes this perhaps to the fact that the (new) faculty members have acquired their culture about the concept of educational technology more broadly than the old faculty members, due to the newness of the era of educational technology and its continuing development, through their study, perhaps for courses in this field or the use of their professors in universities for educational technology and training them during university studies. This is consistent with the results of the study of (Abdul Majeed (2000), Al-Qadri and Al-Shdeifat (2002), which showed differences in the level of awareness in favour of those with short experience, while it differs with the results of Hakami (2010) and Muhammad (2020), which showed no differences in the level of awareness attributed to years of teaching experience.

- Results related to the third question, which states: What are the attitudes of faculty members at Albaydha University towards educational technology? To answer this question, the researcher calculated the arithmetic averages and a T-test for one sample, as well as the standard deviation of it to find out the degree of consensus or disparity of opinions among the study sample individuals towards each paragraph of the measure of the trend towards educational technology, as shown in the following table:

Table 5. The arithmetic averages, deviations, percentages, and "t" values for the paragraphs of the trend scale

#	Item	Mean	Std. Dev.	"t" value	Level of significance	Mean average	Materiality	Level to mean
1	I use educational technology when I'm not convinced of its usefulness	2.92	0.46	8.107	0.000	0.48	58.40	Average
2	I do not like to use educational technology in my teaching, because preparing it requires time, effort and preparation in advance	3.67	0.51	4.052	0.000	0.27	73.40	High
3	I only use educational technology to please my superiors	2.95	0.43	8.128	0.000	0.45	59.00	Average
4	I tend to use educational technology to encourage my students to participate in the topic of the lesson	3.63	0.58	3.109	0.003	0.23	72.60	High
5	Educational technology can be dispensed with in teaching	2.80	0.40	11.522	0.000	0.60	56.00	Average
6	The use of educational technology in teaching leads to the loss of the educational process of its human character	3.73	0.61	4.254	0.000	0.33	74.60	High
7	I do not enjoy using educational technology to teach students	2.92	0.38	9.816	0.000	0.48	58.40	Average

(continued)

Table 5. (*continued*)

#	Item	Mean	Std. Dev.	"t" value	Level of significance	Mean average	Materiality	Level to mean
8	With educational technology, I can bring the outside world into the classroom	3.63	0.52	3.478	0.001	0.23	72.60	High
9	I fear that my students will get messy when I use educational technology	3.02	0.57	5.235	0.000	0.38	60.40	Average
10	I do not tend to use educational technology to teach difficult subjects	2.85	0.52	8.272	0.000	0.55	57.00	Average
11	I think using educational technology in teaching wastes time, effort and money	2.77	0.43	11.502	0.000	0.63	55.40	Average
12	Educational technology helps increase the linguistic wealth of my students	3.70	0.46	5.028	0.000	0.30	74.00	High
13	A successful teacher is one who can communicate information to his students without the help of educational techniques and methods	3.63	0.49	3.719	0.000	0.23	72.60	High
14	It is difficult for educational technology to succeed in contributing to the teaching of humanities and literature	3.70	0.50	4.671	0.000	0.30	74.00	High
15	The expected return from using educational technology in teaching is much less than the costs of obtaining it	3.68	0.60	3.680	0.001	0.28	73.60	High
16	I like to use educational technology in my teaching when I feel that my students are showing boredom	3.58	0.53	2.679	0.010	0.18	71.60	High
17	I see in the scientific disciplines a wide scope for using technology in education	2.85	0.36	11.831	0.000	0.55	57.00	Average
18	I think that the use of educational technology limits the development of the creativity and innovation elements of the learners	3.58	0.62	2.295	0.025	0.18	71.60	High
19	The adoption of educational technology helps me take into account the individual differences between my students	2.98	0.54	6.016	0.000	0.42	59.60	Average
20	I tend to use educational technology because it transforms the teacher's role from a mentor to a mentor	3.68	0.77	2.850	0.006	0.28	73.60	High
21	I believe that a successful teacher is one who adopts the concepts of educational technology in word and deed	3.77	0.47	6.114	0.000	0.37	75.40	High
22	The use of educational technology is a threat to my work as a teacher	3.05	0.59	4.561	0.000	0.05	61.00	Average
	Total degree	3.32	0.26	4.25	0.000	0.08	66.40	Average

It is evident from the data of Table 5 that the arithmetic averages in general for the paragraphs of the trend toward educational technology measure came in average proportions, where the total arithmetic average of the scale paragraphs reached (3.32), and with an agreement on the content of the paragraphs that measure the trend (66.40%), and this confirms the existence of a trend positive average level among the study sample towards educational technology.

Item (21) related to (I believe that a successful teacher is one who adopts the concepts of educational technology in word and deed) ranked first with a high arithmetic average of (3.77), and an approval rating of (75.40%) for the content of this paragraph.

Item (6) related to (the use of educational technology in teaching leads to the loss of the educational process of its human character) came in second place with a high arithmetic average of (3.73), and with approval (74.60%) for the content of this paragraph. Item (11) came in the last place related to (I believe that the use of educational technology in teaching wastes time, effort and money) with an arithmetic average of (2.77) and a percentage of approval (55.40%) for the content of this paragraph.

As for the most consistent paragraphs in the trend towards educational technology among the study sample, item (17) was its standard deviation (0.36), followed by item (7) with a standard deviation of (0.38).

The results of (t) test for one sample (One-Sample Test) came in support of what was reviewed, where the average difference between the items of the trend scale and the approved average (3.40) ranged between (0.05) (0.63) and the (t) values were at this level of the difference is (4.561) and (11.502), respectively, and all of them are statistically significant values at the significance level (0.05).

In general, the average difference between the average score of the overall trend and the value of the approved average of the trend scale was (0.08), and the value of (t) at this level of the difference was (4.25), which is a statistically significant value at the level of significance (0.05), which means that there is a positive trend at an average level among the study sample individuals towards educational technology.

This may be mainly due to the information that reaches the faculty members in various ways, such as the media such as television, radio, newspapers, or from experts, about the possibilities of educational technology and its role in the teaching and learning process to support education and enhance learning and provide the opportunity for both the teacher and the learner to use this technology in the form. The appropriate and to achieve the educational goals.

This result is consistent with the results of the study by Hong et al. (2003), Sharif (2018) and El Gohary (2020), which showed positive trends towards educational technology.

- Results related to the fourth question, which stated: Are there statistically significant differences in the attitudes of faculty members at Albaydha University towards educational technology due to years of teaching experience (less than 5 years - more than 5 years)?

To answer this question, the arithmetic averages and standard deviations of the study sample's attitudes towards educational technology were calculated according to the years of experience variable (less than 5 years - more than 5 years), and the (t) test was used to find out the significance of the differences between the arithmetic averages. The results were as shown in the following table:

Table 6. The arithmetic averages, standard deviations, and the value of the t-test to indicate the differences between the averages of the study sample trends towards educational technology according to the years of experience in education variable

#	To the years of experience in education variable	No.	Mean	Std. Dev.	T value	Level of significance
1	Less than 5 years	23	76.87	3.88	4.624	0.000
2	More than 5 years	37	70.76	5.55		

It is evident from the results of Table 6 that there is an effect of the years of experience in education variable on the level of the trend towards educational technology, where the total arithmetic mean of the level of the study sample who have less than five years' experience reached (76.87), with a standard deviation of (3.88). While the total arithmetic mean of the level of trend of the study sample who have more than five years' experience was (70.76), with a standard deviation of (5.55), and the value of "T" was (4.624), with a level of significance (0.000). This indicates the existence of statistically significant differences at the level of significance ($\alpha = 0.05$) between the mean level of the study sample's attitudes towards educational technology, and in favour of the study sample who have less than five years of experience.

The reason for the high level of trend towards educational technology among faculty members who have less than five years of experience may be attributed to the knowledge component that they possess through their academic study of university courses, as the educational technology subject has become a compulsory requirement for all students. This is in addition to the duties that might have been requested from them through databases and the Internet, which led to the creation of a positive trend towards educational technology in the education process more than the old faculty members who may not have been exposed to studying any courses in the field of educational technology during their university studies.

This differs with the results of the study of Abdul Majeed (2000) and Al-Saeed (2020), which showed that there are no statistically significant differences between the sample members due to the variable of years of experience in education.

- Results related to the fifth question, which states: Is there a total relationship between the awareness of faculty members at Albaydha University about the concept of educational technology and their attitudes towards it?

To answer this question, the Pearson correlation coefficient was calculated between the extent of the study sample's awareness of the concept of educational technology, and their attitudes towards it, as shown in the following table:

Table 7. Pearson correlation coefficient between the average level of awareness about the concept of educational technology and the trend towards it

#	Domain	Total mean	Correlation coefficient	Level of significance
1	Awareness	2.59	0.409	0.001
2	Trend	3.32		

It is noticed from Table 7 that there is a positive and statistically significant relationship between the level of awareness of the concept of educational technology and the level of trend towards it, as the results indicate that the correlation coefficient reached (0.481), and this correlation is positive and statistically significant at the level of significance ($\alpha = 0.001$), which confirms the strength of the relationship between the level of awareness of the concept of educational technology and the trend towards it. Accordingly, it appears that the higher the level of awareness of the concept of educational technology among the faculty members, the more positive attitudes they have towards it. This is consistent with the results of Abdul Majeed (2000) study, which showed a positive relationship between awareness of the concept of educational technology and the trend towards it.

Suggestions:

In light of the findings of this study, the researcher recommends the following:

- Developing the awareness of faculty members about the concept of educational technology, and its role that it can play in the teaching and learning process if it is used properly.
- Conducting training courses to familiarize faculty members with educational technology and train them in its use.
- Work to change the negative attitudes of faculty members towards educational technology, by making them aware of the importance of their role in light of educational technology.

References

Ashteiuh, F.F., Olayan, R.M.: Instructional technology, theory and practice, 1st edn. Safaa House for Publishing and Distribution, Amman, Jordan (2010)

Agha, I., Usthath, M.: Educational research design, 4th edn., Gaza, Palestine (2002)

Al-Jamal, M.H.: The reality of using educational technology and information in learning resource centers in schools in the Kingdom of Bahrain, from the point of view of the learning resource centers' specialists. J. Educ. Psychol. Sci. 6(1), 121–151 (2004)

El Gohary, H.K.A.G.: The awareness of faculty members at Prince Sattam bin Abdulaziz University using the e-learning platform in light of the outbreak of the Covid 19 virus and their attitudes towards inquiry learning by investigation. Arab J. Sci. Res. Dissemination. J. Educ. Psychol. Sci. 4(46), 40–63 (2020). Arab Journal of Science and Research Publishing

Haji Issa, M.: The reality of educational and professional preparation and the reality of United Arab Emirates schools in the field of educational technology. Ajman University's J. Sci. Technol. 6(3), 205–222 (2001)

Hakami, T.T.: The reality of the culture and use of ICT faculty members at Umm Al-Qura University in teaching, an unpublished master's thesis, Umm Al-Qura University, Saudi Arabia (2010)

Salem, A.M.: Education and e-learning technology. First Edition. Al-Rashed Library Publishers, Riyadh (2004)

Salem, A.M., Saraya, A.: Education Technology System. 1st floor, Al-Rashed Library, Riyadh, Saudi Arabia (2003)

Al-Saeed, B.A.-B.A.: Attitudes of faculty members at Jazan University towards employing electronic learning tools, the blackboard platform in the educational process in line with the implications of the Coronavirus, the Arab Journal of Science and Research Publishing. J Educ. Psychol. Sci. 4(7), 1–19 (2020)

Salama, A.H.: Communication and education technology. I 1. Al-Yazuri Scientific House, Amman, Jordan (2001)

Salama, A.H.: Communication and education technology. Al-Yazuri Scientific House, Amman, Jordan (2007)

Al-Shaer, M.F.M.: The extent to which students of home economics are aware of employing the latest educational and information technology and their motivation for achievement. Arab Res. J. Fields Specific Educ. (Seventeenth Issue), 349–381 (2020)

Sharif, B.b.N.M.: The extent of awareness of digital and smart educational technologies for faculty members in Saudi universities and their attitudes towards them. J. College Educ. Al-Azhar Univ. (179), Part 1, 601–650 (2018)

Shami, N.S., Ismail, S.S., Muhammad, M.A.S.: Introduction to educational technology. First floor, Dar Al Fikr, Amman, Jordan (2008)

Abdul Majeed, M.M.: The extent to which science teachers are aware of educational technology innovations and their trends towards employing them. In: Fourth Scientific Conference (Scientific Education for All), from July 31 to August 13, vol. 1, pp. 309–338. Egyptian Association for Scientific Education (2000)

Ghazzawi, M.Y.: Educational technology between theory and practice. The Modern World of Books, Irbid (2007)

Al-Ghishan, R.I.: The degree of interest of primary school teachers in public schools in Amman's education directorates in education technology, and the students' attitudes towards it. Ph.D. thesis, University of Jordan, Amman, Jordan (2005)

Al-Qadiri, S.A., Shdeifat, Y.: The level of education attained in the educational computer of managers and teachers working in the Northern Badia Education Directorate. Damascus Univ. J. 18(2), 149–184 (2002)

Al-Qahtani, A.S., Al-Muaither, R.A.: The awareness of the faculty members at Princess Noura University with stereoscopic imaging technology (hologram) in distance education. Arab Stud. Educ. Psychol. (71), 299–333 (2016)

Muhammad, A.H.A.: The extent of awareness of teachers with special needs about the importance of e-learning in the Sultanate of Oman. J. Humanit. Soc. Sci. Gener. (66), 151–168 (2020)

Hong, K., Ridzuan, A., Kuek, M.: Students' attitudes toward the intern for learning: a study at a University in Malaysia. Educ. Technol. Soc. 6(2), 45–49 (2003)

Clark, K.D.: Urban middle school teachers' use of instructional technology. J. Res. Comput. Educ. 33(2), 178–195 (2000)

Labusch, A., Eickelmann, B.: Computational thinking and problem-solving – a research approach in the context of ICLLS 2018. In: Langran, E., Borup, J. (eds.) Proceedings of Society for Information Technology & Teacher Education International Conference, pp. 3724–3729 (2018)

Sultan, M.A.: The need to go beyond "Techno Centrism" in educational technology: implementing the electronic classroom in the Arab world. In: Billeh, V., Mawgood, E. (eds.) Educational Development through Utilization of Technology, pp. 165–173. UNESCO Regional Office for Education in the Arab State, Beirut (2001)

Author Index

Abbood, Elaf Ali   49
Abdulkhaleq, Ali Hussein   63
Abdulkhudhur, Hanan Najm   131
Ahmed, Zainab J.   17
Al Hassani, Safaa Alwan   63
Al-Abrat, Nayef Ali Saleh   189
Al-Abyadh, Mohammed Hasan Ali   189
Al-Hammad, Ahmed A.   33
Alheeti, Khattab M. Ali   111
Alhussein, Duaa Abd   163
AlKhayyat, Atheel N.   33
Al-Mamory, Safaa O.   176
Al-Noori, Ahmed H. Y.   33
Al-Zaydi, Zeyad Qasim   131

Couturier, Raphael   176

EL Abbadi, Nidhal K.   63

George, Loay E.   17, 87

Habeeb, Imad Qasim   131
Hadi, Asaad Sabah   3
Hadi, Hend A.   87
Hamza, Weam Saadi   99
Harb, Hassan   163
Hasan, Ayat S.   99
Hassan, Enas Kh.   87

Ibrahim, Hassan M.   99
Idrees, Ali Kadhum   163, 176
Idrees, Sara Kadhum   176
Ismail, Qunoot Mustafa   49

Jasim, Mahdi Nsaif   147

Kareem, Aythem Khairi   111

Najem, Yasir Abdalhamed   3
Neamah, Rusul Mohammed   49
Noman, Haeeder Munther   147

Shyaa, Methaq A.   99

Printed in the United States
by Baker & Taylor Publisher Services